More Than Us

Discover How We Are Going To Find

Life In The Universe

George Davis.

© **Copyright 2018 - All rights reserved.**

This document is geared towards providing exact and reliable information in regards to the topic and issue covered. The publication is sold with the idea that the publisher is not required to render accounting, officially permitted, or otherwise, qualified services. If advice is necessary, legal or professional, a practiced individual in the profession should be ordered.

- From a Declaration of Principles which was accepted and approved equally by a Committee of the American Bar Association and a Committee of Publishers and Associations.

In no way is it legal to reproduce, duplicate, or transmit any part of this document in either electronic means or in printed format. Recording of this publication is strictly prohibited and any storage of this document is not allowed unless with written permission from the publisher. All rights reserved.

The information provided herein is stated to be truthful and consistent, in that any liability, in terms of inattention or otherwise, by any usage or abuse of any policies, processes, or directions contained within is the solitary and utter responsibility of the recipient reader. Under no circumstances will any legal responsibility or blame be held against the publisher for any reparation, damages, or monetary loss due to the information herein, either directly or indirectly.

Respective authors own all copyrights not held by the publisher.

The information herein is offered for informational purposes solely, and is universal as so. The presentation of the information is without contract or any type of guarantee assurance.

The trademarks that are used are without any consent, and the publication of the trademark is without permission or backing by the trademark owner. All trademarks and brands within this book are for clarifying purposes only and are the owned by the owners themselves, not affiliated with this document.

TABLE OF CONTENTS

Prologue – Are we alone?	1
What is SETI?	3
Size of the Universe	5
Forms of life	10
Life as we know it	10
Life as we don't know it	11
Telescopes	13
Factors in the search for life	16
Where should we look for signals?	18
Signals from extraterrestrial civilizations	18
Signs of life in general	47
Conclusion on where we should search	68
Should we send signals ourselves?	71
METI	71
Past attempts	75
San Marino Scale	75
Why haven't we seen them?	77
Category I – we are rare	78
Category II – we are alone	80
Category III – we are one among many	81
What happens if we find something?	88
Confirmation of a signal	88
Post-detection cover-up	91
Cultural impact of extraterrestrial contact	93
Projects	97
What have we done so far?	97
What are we doing right now?	105

How can I get involved?	118
Epilogue – So, are we alone?	122
Acknowledgments	123
Interesting resources	124
References	129

Prologue – Are we alone?

Are we alone? That's an ancient question. One that has kept many people awake at night. One that is still unanswered.

Extraterrestrial life, whether intelligent or not, still awaits discovery. If there is any.

With all the exciting findings of planets around other stars and lifeforms that can live under conditions we previously thought impossible, it seems very likely that life has arisen elsewhere in the universe. But in fact, we still don't know. There are still too many unknowns.

I am optimistic about the chances of finding other life, even intelligent life, somewhere in the cosmos. You probably are too. At least, you're obviously pondering this question.

However, so far we have no tangible evidence of life elsewhere in the universe. We don't even know enough to realistically estimate the probability.

The fact that we haven't found anything yet should not discourage you. Humans have thought about this question for millennia. Only recently have we developed the technology to not just think about it, but to actually search for life in the universe.

Systematic scientific searches have started no earlier than the second half of the 20th century. And we have, due to technical and financial limitations, just been able to search a little bit of the sky for a little bit of time.

I like an analogy I heard from Dr. Jill Tarter. It goes something like this: Concluding from the research done so far that there is no life in the universe is like taking a glass of water out of the ocean and concluding that there is no life in the oceans because there is no fish in the glass.

This book will give you a broad overview of the search for life in the universe. We will, among other things, cover topics like different forms of life and where to search for them. We will discuss the questions if we should send signals ourselves and how it is possible that we haven't heard from anybody yet. We will also take a look at the efforts that have been made to find life in the cosmos and what happens if a signal is detected.

What is SETI?

What exactly is *SETI?*

SETI is an acronym and stands for the *Search for ExtraTerrestrial Intelligence.*

There is no one SETI, nor is there a single institution that does all things SETI. There are, in fact, many different institutions and people involved in this search. Some of them, as well as past and present projects in the field of SETI, can be found in the following pages.

In essence, SETI is a collective term for all scientific efforts to look for intelligent life in the universe that does not currently live on planet Earth. Hence the name extraterrestrial intelligence.

Note the word "scientific." SETI describes efforts that use scientific tools to attain evidence in the form of data which can be confirmed by other people through their own observations and calculations. It is not about claimed alien abductions in the backyard or circles in cornfields. We'll take a quick look at this later in the book.

There are more or less popular subcategories like radio SETI, optical SETI or SETA. We will cover them in more detail soon.

As Dr. Jill Tarter, a famous researcher in the field, pointed out, SETI is actually a search for extraterrestrial technology. That's because we are looking for artificial signals that are clearly made by technology. From that, we infer that to build this technology the extraterrestrials have to be intelligent.

SETI usually doesn't include the search for other life forms that are not intelligent. Looking for microbial life on Mars, for instance, is often not considered to be part of SETI. However, the boundaries are a bit blurry.

In this book we will take a broad approach and not only look at intelligent life, but life in general.

Size of the Universe

To get an understanding of how large a task SETI is, let's first take a look at the size of the universe. The universe is huge. Let's see just how huge it is.

We begin our journey on planet Earth, our beautiful home. The Earth has a diameter of 12,742 km (7,917 miles). Not bad compared to the sizes we see in our everyday lives. But we're leaving the realm of daily life now and go on to astronomical scales.

Our first stop is the Moon. It's smaller than the Earth with a diameter of just 3,474 km (2,159 miles). But it's already 384.400 km (about 238,900 miles) away. If that sounds far away, wait until we look at our Sun.

The Sun is 150 million km (93 million miles) away from Earth. Okay, now our everyday life units of distance start to get impractical. So let's introduce another unit, namely the Astronomical Unit (AU). It is the average distance between the Earth and the Sun. So, as we've just seen, it's about 150 million kilometers (93 million miles). Now we're equipped to move on.

If we turn outward from the sun now, we see the last planet in our solar system, Neptune. In case you have missed it: Pluto is no longer officially considered a planet. So that makes Neptune the last planet in our solar system. Neptune is 30 AU away from the Sun. That means it's 30 times farther away from the Sun than the Earth is from the Sun.

Our next step towards the edges of the universe is the outer limit of our solar system. How big is the solar system? Well, that depends on your definition. One way to define it is by the Heliopause (the region where the solar wind can't keep up against the pressure of the interstellar medium anymore). This is the definition *NASA* used when they announced that *Voyager 1* had left the solar system. In this case, the outer limit of the solar system would be 122 AU from the Sun.

Another way to look at it is by the gravitational influence of the Sun. According to this definition, we would place the boundary of the solar system at a distance where the gravity of the Sun is still strong enough to capture objects. In other words, objects at this distance would still barely orbit the Sun rather than anything else. This limit is estimated to be at around 125,000 AU from the Sun. That's more than 1,000 times as far out as the Heliopause! As you can see, the size of our solar system is a matter of perspective, and the distance from the Sun at which it ends varies greatly. There are other definitions as well, but for now, let's say the end of our solar system is somewhere between 120 to 125,000 AU from the Sun.

At these scales, even the AU slowly begins to get impractical. Therefore let us take a look at yet another unit of distance widely used in astronomy. It is the light-year. A light-year, unsurprisingly, is the distance that light can travel in one year. Although very fast, light doesn't move at infinite speed. Instead, it moves at the aptly named speed of light, which is roughly 300,000 km/s. This means in one year light could travel 9.5 trillion kilometers (5.9 trillion miles) or approximately 63,000 AU.

Do you think a light-year is an enormous distance? Well, even our nearest star, Proxima Centauri, is 4.243 light-years away.

Our next stop is the Milky Way Galaxy. It is home to our own Sun, Proxima Centauri and somewhere between 100 billion and 400 billion other stars (it's not that easy to tell, because we are in the Milky Way and not just looking at it from the outside). And there are not only a lot of stars here. It is currently estimated that our galaxy is also home to at least 100 billion planets. The Milky Way was long thought to have a diameter of 100,000 light-years. But new research from recent years suggests that it could actually be up to 200,000 light-years across. We are already talking about immense distances and an incredible amount of stars and planets. But we're not nearly at the end of our journey. Let us continue.

The nearest major galaxy from the Milky Way is the Andromeda Galaxy. It has a diameter of 220,000 light-years and contains about 1 trillion stars. Its distance from Earth is 2.5 million light-years. The light that we see when we look through a telescope at the Andromeda Galaxy traveled 2.5 million years to get here!

Now we're getting to even more incomprehensible scales. The Milky Way and Andromeda are part of a set of 54 galaxies, called the Local Group. It measures 10 million light-years across.

Depending on your definition, we then get to the Virgo Supercluster or the even bigger Laniakea Supercluster (of which Virgo is only a small part). And it just gets bigger and bigger until we reach the end of the known universe.

It might surprise you to hear that the real size of the universe is unknown. That's because we cannot observe all of it. The diameter of the observable universe is about 93 billion light-years. We cannot see regions that are farther away from us simply because the light from these regions didn't have enough time to reach us since the beginning of the universe.

I told you the universe is huge.

The immense size of the universe and the limited speed at which light can travel have interesting consequences for the search for life in the cosmos.

Moving almost 300,000 km every single second sounds fast, but compared to the size of the universe it seems not that fast anymore. The distance to even our nearest star (Proxima Centauri) is so big that it takes its light more than four years to reach us. Therefore, every time we look up at the stars we look back in time. We see our nearest star Proxima Centauri not as it is now, but as it was more than four years ago. We don't even see our own star, the Sun, as it looks right now, but as it looked about 8 minutes ago. That's how long it takes light from the Sun to reach us here on Earth.

The limit on the speed with which light can travel is a problem for SETI or the search for life in general. There might be a civilization right now in our galaxy, let's say 50,000 light-years away from Earth. The signals we receive now from their star system are 50,000 years old. So, we might not be able to see anything of their existence although they exist right now.

This is, of course, also a problem for communicating with such a civilization. Let's say we send a message to this civilization 50,000 light-years away from us. It would take 100,000 years before we would get an answer, as the signal has to travel 50,000 years to reach them and then their response has to travel another 50,000 years to get to us.

A civilization that is more advanced than we are might, of course, have found a way to communicate faster. Right now we have no idea how that could work, but that doesn't mean it's impossible.

Forms of life

From our human perspective, two forms of life might be out there in space: *Life as we know it* and *life as we don't know it*.

Life as we know it

Life as we know it is not clearly defined, but generally, it means life as we find it here on Earth. There is, of course, a broad spectrum of life forms here on Earth. But all of them are based on carbon and need liquid water.

In recent years, we learned a lot about so-called "extremophiles." These are life forms found here on Earth that can not just survive but thrive in conditions that seem extreme from a human perspective (hence the name). But for these life forms, it's just a cozy environment to live in. We've found life in wet environments and dry ones. We've seen it in acidic and alkaline environments. We've found it where it's hot and where it's cold. We've even found it in high radiation environments in nuclear power plants. But all of these life forms, from humans to dolphins to extremophiles are based on carbon and require water in its liquid form to exist.

Life as we know it is what we are mostly looking for when we search for other life out there in the universe. That's simply because we know what to look for. We understand what life forms like this need (namely liquid water) and how to detect them because our instruments are constructed to find life like we found here on Earth.

Life as we don't know it

It's a lot harder to find something when you don't know what to look for. But that doesn't mean it doesn't exist. *Life as we don't know it* then is every form of life that is not in the category of *life as we know it*.

Life like this is theoretically possible although we've never seen anything like it. All life on Earth is based on carbon, but it is possible that there is life based on other atoms than carbon. Silicon is a hot candidate as it has many similarities with carbon. Also, it is an abundant element in the cosmos.

Liquid water might be replaceable as well. Other liquids could, in principle, take the role water has for life on Earth, among them ammonia and methane.

We still don't know how life emerges. Therefore, other forms of life are mere speculation at this point. We know that *life as we know it* can exist because we have clear evidence of its existence here on Earth. But if forms of life other than what we know are indeed possible, chances are it did evolve somewhere in the universe.

And we might not even have to look very far. The hottest candidate for *life as we don't know it* in our solar system is Saturn's moon Titan. But we'll get to that later.

The problem is, scientists are not sure how to detect *life as we don't know it*. But this is a topic that has been much discussed in recent years. So, who knows? We might find life in our solar system even at places we thought impossible some years ago. If life based on other elements than life here on Earth is indeed possible, that would definitely increase the likelihood of finding life elsewhere dramatically.

Telescopes

As we shall soon see in more detail, every form of electromagnetic radiation is essentially the same. The reason there are different parts of the spectrum with different names (such as radio waves, visible light, infrared) is that these portions of the spectrum where discovered separately from one another. And that is because no one instrument can detect all forms of electromagnetic radiation. You need a different device for every part of the spectrum (although there are some overlaps).

The kind of radiation that was discovered first (and for obvious reasons) is visible light. These are the frequencies that our eyes can see. So, obviously, it was the first kind of radiation that we observed from the stars. Even our ancestors living in caves could watch the stars in the visible part of the spectrum. In fact, this frequency range is the only one we can observe without a specialized (technological) instrument.

Well, you could detect infrared light with your skin (as heat), but you would never be able to observe objects in space that way (apart from our own sun, the heat of which you can clearly feel on your skin on a sunny day). You cannot even observe objects in space with a specialized infrared telescope if it sits here on the ground. That's because of our atmosphere. And this is the reason infrared telescopes are usually in space (there is one exception, *SOFIA*, which is actually a telescope in an airplane).

We can observe visible light with our eyes. But the problem is the collecting area. With your eyes, you are not able to see more than 2,000 - 3,000 stars in the night sky. And you have to be in a really dark place to see even this number of stars. Everything else you will virtually never be able to see with your eyes because the objects are too faint. And that is the reason why we build optical telescopes despite the fact that we can see visible light even without any technological instrument. With optical telescopes, it's a matter of bigger is better. A lot of people think that optical telescopes are primarily for magnifying objects. But that is not the case. The main reason we build ever bigger optical telescopes is that the objects we want to observe are very faint. And the bigger the telescope, the more light it can gather. And that way, it can see objects that are just not shining very bright or that are much farther away and therefore appear dimmer, because less of their light reaches us.

To illustrate this, imagine a lightbulb. When it is in your living room right above your head, it is very bright. But if you go down the street and look at it from a hundred meters away, it will appear a lot dimmer, although it is still emitting the same amount of light it did when you stood right under it. And that is because objects appear fainter the farther you are away from them. There are various reasons for this dimming, and it applies to any form of electromagnetic radiation, although not to all of them equally.

You sure have seen (at least on pictures) what telescopes look like that are used to examine the radio and microwave parts of the spectrum. Radio telescopes are (usually) these huge dishes that look a lot like the satellite dish you use to receive your TV signals. And they function in much the same way. This form of radio telescope with which nearly everyone is familiar is, however, not the only form of radio telescope. You have undoubtedly seen another way to receive radio waves already. Think of your car. It probably has an FM radio with which you can hear different radio stations. So, apparently, your vehicle can receive radio waves. But it usually has nothing close to a satellite dish on the roof. There must be more than one kind of antenna that can receive radio waves. And there is, indeed. The difference between different shapes of radio antennas is the frequencies that they can detect. Yes, you need different kinds of telescopes to catch the whole radio part of the spectrum.

And then there are specialized telescopes for detecting x-rays and gamma rays. And yet other kinds of telescopes to discover other things that don't belong anywhere on the electromagnetic spectrum, like neutrinos and gravitational waves.

As you can see, we need a lot of fancy technology and lots of different instruments to discover all that is out there. Regarding the exploration of the universe, there's definitely more than meets the eye.

Interspersed in the following pages, you'll find a few of the telescopes used for the search for extraterrestrial life in the universe. This selection is by no means complete. These are most definitely not all telescopes that were or are used for SETI or, more generally, the search for extraterrestrial life. But as this is not a book about telescopes, I've just chosen a few examples to show you some of the amazing instruments that humans have come up with to discover (and solve) the mysteries of the universe.

Factors in the search for life

In November of 1961, the first scientific conference solely on the (modern) Search for Extraterrestrial Intelligence was held at the *Green Bank Observatory* in West Virginia, USA.

Ten people attended this conference. They were Dana Atchley, Melvin Calvin (who won the Nobel Prize during the meeting), Frank Drake, Su-Shu Huang, John C. Lilly, Philip Morrison, Barney Oliver (a name you will often hear in the SETI community), J. Peter Pearman (the one who came up with the idea for the conference), Carl Sagan (a name you will hear a lot in all of astronomy) and Otto Struve (who served as the host).

In honor of Lilly's work on dolphin communication (which was controversial at the time) they called themselves "The Order of the Dolphin."

During the preparations for the meeting, Frank Drake realized that they needed an agenda. He wrote down all the factors affecting the likelihood that we can detect communication from ETIs (Extraterrestrial Intelligence). He put these topics in a nice equation format. The result, now known as the *Drake equation*, has become very popular.

The *Drake equation* is still a useful tool today. Right now, it is not that useful for actually calculating the number of communicating civilizations in our galaxy because of the significant uncertainty of at least some of the factors. However, it is a good overview of what factors play a role in detecting ET life. And it is just fun to play with different values and see what number of communicating civilizations you get in different scenarios. As our knowledge increases and we get a clearer understanding of the yet very uncertain variables in the *Drake equation*, the results will, of course, get more and more realistic.

The *Drake equation* looks as follows:

$$N = R_* \cdot f_p \cdot n_e \cdot f_l \cdot f_i \cdot f_c \cdot L$$

where:

N = the number of civilizations in our galaxy with which communication might be possible

and

R_* = the average rate of star formation in our galaxy
f_p = the fraction of those stars that have planets
n_e = the average number of planets capable of supporting life around these stars
f_l = the fraction of planets capable of supporting life that actually develop life
f_i = the fraction of planets with life that develop intelligence
f_c = the fraction of intelligent life that develop technology and transmit signals that can be detected by others
L = the lifetime of such a civilization

Where should we look for signals?

We want to find life elsewhere in the universe. But how do we find it? Where should we look?

Signals from extraterrestrial civilizations

We've already seen that the universe is vast and that there are different forms of life to look for. As if these factors weren't enough to make SETI a big task, there is also a whole lot of different ways that ETIs could use to communicate. Only if we use the right method to listen to their signals will we ever be able to hear them. Let's now take a look at all the options we have.

Electromagnetic spectrum

There is one form of electromagnetic radiation that you are undoubtedly familiar with, and that is visible light. However, this is only a small part of the whole electromagnetic spectrum.

All electromagnetic radiation is essentially the same. Different forms of electromagnetic radiation are produced by different processes and measured with different instruments. But there is no fundamental difference between the parts of the spectrum. In fact, the names and boundaries of the parts are entirely arbitrary. The only exception is visible light, which is obviously electromagnetic radiation that our eyes can see.

The electromagnetic spectrum consists of (in order of increasing frequency):
- radio waves (AM/FM radio; television)
- microwaves (microwave oven; radar)
- infrared (TV remote control; night vision goggles)
- visible light (light our eyes can see)
- ultraviolet (UV light from the sun that tans your skin)
- X-rays (airport security scanners; used by your doctor to see if you broke one of your bones)
- Gamma rays (used by doctors to kill cancer cells; can also lead to acute radiation syndrome, aka radiation sickness, depending on the dose)

Note that all of the above-listed applications are terrestrial, but natural processes in the universe also produce all of these types of radiation. That's why we use various types of telescopes to see the universe in all its glory. Every part of the spectrum tells us different things about the cosmos, its structure and the processes happening within it. There are lots of things we wouldn't be able to see and therefore would know nothing about if we just used our eyes (or optical telescopes) to look at the stars.

It is a continuous spectrum. The different names for the parts (also called bands) of the spectrum (radio waves, Gamma rays, etc.) have historical reasons as they were discovered separately. There are also no clearly defined boundaries between the different bands.

All parts of the spectrum consist of electromagnetic waves. The only difference is their frequency, wavelength and the energy they carry. These properties are all linked together.

The wavelength of an electromagnetic wave is the distance over which the wave's shape repeats. So, for example, the length from one crest to the next.

The frequency, on the other hand, describes how often that happens over a given period of time. The typical unit of frequency is the Hz (Hertz, named after Heinrich Hertz). The Hz specifies how often the wave shape repeats over the time of one second.

Because electromagnetic waves travel with the speed of light, the wavelength is inversely proportional to the frequency. In other words: the smaller the wavelength, the higher the frequency.

The energy of a wave is proportional to its frequency. The higher the frequency, the higher the energy carried.

Some parts of the spectrum can't get through the Earth's atmosphere. That's why we put some of our telescopes in space. We wouldn't be able to see that kind of radiation from the ground at all or at least not nearly as good. Also, even for types of radiation that do reach the ground, the quality of the image is usually better when taken from space because there is no atmosphere in the way and generally less interference (e.g., from terrestrial radio sources). But of course, it costs way more to put a telescope in space than to build it on the ground. That's why we have to choose which instrument to put up there carefully.

This kind of radiation was first observed in 1951. The frequency falls into the microwave part of the electromagnetic spectrum and therefore belongs in the realm of radio astronomy. It is a frequency that is now regularly observed by radio astronomers because it can easily penetrate clouds of interstellar dust which visible light cannot. Radiation with the frequency of 1,420 MHz was actually used to "see" the spiral structure of our galaxy for the first time.

Cocconi and Morrison favored the radio part of the spectrum from 1 GHz to 10 GHz because such signals are relatively easy and energy efficient to produce. Then they reasoned that, within this range, the frequency of 1,420 MHz would be the most promising to look for signals from ETIs. Radio waves of this frequency are not attenuated much in space or the Earth's atmosphere. Furthermore, they reasoned that this frequency must be known to every observer in the universe because every technologically capable civilization would discover it relatively soon after gaining the capability of detecting radio waves.

They also assumed that an extraterrestrial civilization would be willing to let newly (technologically) evolved civilizations know that someone else is out there. Therefore an ETI would use a technology and frequency that would be detectable by another civilization at an early stage of their technological development.

At least in the case of humanity, this is true. We were able to detect such signals at an early stage. The first astronomical radio source was discovered by Karl Jansky in the early 1930s. Around the middle of the 20th century, it was predicted that neutral hydrogen could produce radiation at 1,420 MHz and shortly thereafter, it was observed. The middle of the 20th century is pretty early in our development of (modern) technology. Around that time computers used vacuum tubes and weight tons (e.g., the *ENIAC* had 18,000 vacuum tubes and weighed 30 tons, had a high-speed memory of about 80 bytes, and it could do 5000 calculations per second).

There is a reason why they proposed to search for signals near a frequency of 1,420 MHz and not just precisely at 1,420 MHz. It's because they knew, even if the source transmitted a signal at 1,420 MHz, it would be slightly shifted when it arrives at a radio telescope on Earth. This is due to the Doppler shift because the source will very likely be moving relative to Earth. Their planet will be revolving around their star. Additionally, their solar system will be revolving around the center of the galaxy. This means that the source of the signal will very likely be moving away from us or towards us. And that leads to a Doppler shift. So an incoming signal would have a slightly lower or higher frequency than 1,420 MHz.

Doppler shift

The Doppler shift (or Doppler effect) is the change of frequency of a received signal due to the motion of the source or the receiver.

What this means is, when the source of a signal or the receiver move in relation to each other, the frequency of the signal changes.

A popular example is a car horn or an ambulance siren. When the car moves towards you, you hear a higher sound than when it has passed you by and is driving away from you.

The reason for this effect is that when the source of the signal moves away from you while sending out the signal, the waves are pulled out a bit because the source is a little bit farther away from you each time a new wave is emitted. This leads to a lower frequency. Conversely, when the source is moving towards you, the waves get compressed together, which increases the frequency.

This effect can be observed for any electromagnetic wave and, as seen in the example above, even for sound waves.

Project Cyclops

In 1971 *NASA* founded a study to investigate how SETI searches should be undertaken. This project, led by Bernard Oliver and John Billingham, was called *Project Cyclops*. The project team came to a similar conclusion than Cocconi and Morrison. But rather than limit the search window to 1,420 MHz they proposed searching in a broader region of the spectrum.

This region is the frequency range between 1,420 MHz and 1,666 MHz, which has been called the *water hole* by Bernard Oliver because its boundaries are the spectral lines of neutral hydrogen and hydroxyl respectively. Together they form water, which should be known to every technologically advanced civilization. Additionally, this region of the spectrum is relatively quiet (not much background noise). All of this makes it a good place to look for signals from ETIs.

They also designed a telescope for observing such signals. As the proposed frequency range is in the microwave part of the spectrum, the proposed instrument was a radio telescope array. With a price tag of $6-10 billion (in 1971), it comes as no surprise that this telescope array was never built. For comparison, that's $37-62 billion in 2018 dollars. This comparison is just for clarifying the immensity of the costs. If this array were to be built today, it would, of course, use more advanced technology. So the original costs would have to be reviewed.

It was actually suggested to begin with a small number of radio dishes and build more and more if no signal was detected. But it seems as though all the people responsible for distributing the money saw was the price tag of the whole array.

The conclusions of the project were published in the *Cyclops Report*. It is in the public domain. You can download it for free in *NASA's* archive[1].

Although the proposed telescope was never built, the report and its conclusions have considerable influence on the field of SETI to this day. It's probably safe to say that every major player in the field of SETI has read this report. It is even sometimes called the "SETI bible."

Some researchers (among them Stuart Kingsley, a popular proponent of Optical SETI) heavily critiqued the report or at least its enormously influential position in the SETI community, because it supposedly led to a sole focus on microwave searches. It is at least undoubtedly true that the *Cyclops Report* influenced much of the SETI work that followed.

Although technology has advanced, the frequencies proposed by Cocconi and Morrison and later the *Cyclops Report* are still a part of most SETI programs today.

Other radio wave frequencies

Today's SETI programs, although they still include the *water hole*, are looking at a much broader set of frequencies. The technological advancements since 1959 make it possible to look not at one radio channel but billions of them simultaneously. Current search programs scan the frequency range from 0.4 GHz up to 10 GHz, the specific range, of course, depending on the respective program.

Optical frequencies

The whole idea of modern SETI started with the publication of the *Nature* article by Cocconi and Morrison in 1959. And although past searches have included more and more channels besides the "magic" frequency of 1.42 GHz (1,420 MHz), they were still almost all confined to the microwave part of the spectrum.

And that is likely because of the hugely influential position of the article by Cocconi and Morrison and the *Cyclops Report* in the SETI community. Both clearly proposed the microwave part of the spectrum as the most likely region for transmissions from ETI and therefore the most favorable place to search. Some of the arguments brought forth by the authors of these two papers are still valid. Others were seriously influenced by the technological state of humanity during their writing.

Technological advancements of the last several years have made it clear that there are other parts of the spectrum suitable for interstellar communication. And microwaves might not even be the best way of communicating. One of these alternatives is visible light.

Now, ETIs would probably not turn a gigantic lightbulb on for us to see. Instead of using a lightbulb that emits in every direction, they would more likely use something that is way more energy efficient. And that is a laser (the acronym for Light Amplification by Stimulated Emission of Radiation). A laser emits a focused beam of light, instead of emitting light in every direction like a lightbulb. That means you need a lot less energy to be seen by your target. But it also means you need to pick a target to focus your laser on.

Technological advancements in the recent years have shown that interstellar communication via optical lasers might be a pretty good idea. But the idea to search for laser signals instead of microwave signals is way older. In fact, it is almost as old as the microwave approach.

In yet another article in the magazine *Nature*, this time in 1961, two years after the article by Cocconi and Morrison and just one year after the invention of the laser, R. N. Schwartz and Charles Townes suggested searching in the optical part of the spectrum. At that time, however, the laser technology was in its infancy (as mentioned above, the laser was invented just one year before the publication of the article). Radio technology, although also not very advanced by today's standards, was way ahead of the laser technology.

So, two of the most important papers in the history of SETI, the article by Cocconi and Morrison and the *Cyclops Report*, clearly favored radio waves as a means of interstellar communication. Furthermore, the radio technology was more advanced than the laser technology. These seem to be the likely reasons why optical SETI was pretty much neglected until the 1990s (at least in the US, there were some attempts in Russia).

Since then, there have been several projects, some of them ongoing, at optical or infrared wavelengths. The latest of them are

- *Breakthrough Listen* (which, in addition to searching for radio waves, is also searching for laser pulses using the *Lick Observatory*),
- the *SETI Institute's All-Sky All-the-Time Search* (which is, at the time of this writing, not yet operational but has already received significant funding), and
- *METI International's* Optical SETI program.

Microwaves have their advantages. Furthermore, the "magic frequencies" around the *water hole* have some significance beyond mere technological advantages. But optical signals also have their benefits. Some of them are:

Visible light (and their surroundings in the electromagnetic spectrum, namely infrared and ultraviolet) have a much higher frequency than microwaves and can consequently send information much faster. If you want to send a lot of information, you can do it a lot quicker with a laser signal.

If a laser pulse is detected, the probability that it is an ETI signal is higher than if we discover a pulse in the microwave part of the spectrum. That's because of interference. A radio telescope can receive a microwave pulse that, at first glance, looks like an interesting signal. But it could have originated from things like spark plugs from nearby cars. Or when someone in the cafeteria opens the microwave oven before it is finished (this actually happened). Interference from terrestrial sources (or our satellites) is such a big deal in radio SETI that it has its

29

own acronym: RFI (Radio Frequency Interference). In the optical part, there is no such terrestrial interference, at least not on the nanosecond (one billionth of a second) timescale that most programs are looking at. Furthermore, as far as we know, there is no natural source of nanosecond laser pulses in the universe.

Compared to radio signals, the devices you need for sending and receiving a message are a lot smaller. A laser pulse would probably not be a faint signal that needs enormous telescopes to receive. A laser with a plausible amount of energy would outshine its host star by thousands, tens of thousands or even hundreds of thousands of times even in a relatively small optical telescope. That is, provided it has the right equipment to detect such pulses (you wouldn't recognize such a pulse with your backyard telescope if you were just looking through it with your eyes). We're not talking about a 100m-telescope here. Instead, a 1m-telescope would suffice.

Another advantage is that with optical, you don't have to choose a specific frequency to listen to. With radio waves, the radio telescope receives a lot of frequencies. But you have to decide which ones you analyze. In the optical range, you don't necessarily have to do that. Due to technical differences, you can set up an optical telescope that can detect a pulse in any frequency within the visible part of the spectrum. So you don't have to choose beforehand at what specific frequencies you want to look. With increases in processing power over the last years we can analyze a lot more frequencies simultaneously in the radio part of the spectrum now, than was the case at the beginning of SETI. But even with the dramatic increase in processing power, the amount of data gathered with radio telescopes is too enormous to be analyzed very thoroughly. Let's take the

Breakthrough Listen program as an example (more on this program later). At the *Green Bank Telescope* alone we're talking about 24 GB of data per second that is recorded during *Breakthrough Listen* observations at the time of this writing. And that is expected to increase in the near future. Even with today's computing power, we cannot look very deep into all of this data in real-time. There might be an incoming signal that we don't recognize when it hits our telescopes.

Luckily, the information is recorded and open to anyone for further analysis. Who knows? An ETI signal might already be recorded and buried deep within these enormous amounts of data, just waiting to be found. So, at least for now, not having to choose a frequency beforehand is still an advantage of optical SETI.

Other parts of the spectrum

To my knowledge, all serious scientific SETI programs took place in the radio, microwave or optical parts of the spectrum (as far as active observations go, not the analysis of archival data). There were/are some attempts in the infrared and ultraviolet. But these are generally considered to be part of optical SETI.

That leaves X-Rays and Gamma Rays. Most scientists consider the use of these parts of the spectrum unlikely for interstellar communication, mainly because of the high amounts of energy necessary to create such signals. But there is no fundamental physical reason that these parts of the spectrum could not be used by ETIs.

It has been proposed that the phenomenon of GRBs (Gamma Ray Bursts) could be an ETI signal created through matter-antimatter annihilation to draw attention. That seems possible but pretty unlikely. The probability of GRBs originating from a natural process is extremely high, and the majority of scientists believe they are of natural origin. Furthermore, even if some of them could be of artificial origin, why use a signal that looks just like a naturally occurring phenomenon that happens regularly in a random spot of the sky? Regardless of where you look in the sky, you will see a GRB at some time. So, if some of these bursts were of artificial origin, that doesn't seem to be a very efficient way to get our attention.

It has also been suggested that advanced extraterrestrial civilizations might be able to modify natural X-ray sources (X-ray binaries in particular) to transmit information[2]. By piggybacking on existing (or planned) X-ray telescopes, these kinds of SETI observations could be made at relatively low cost. To my knowledge, although proposed by Robin Corbet[3] in the 1990s, there have been no SETI observations in the X-ray part of the spectrum, neither with specialized telescopes for SETI nor have there been any piggyback SETI devices installed on X-ray telescopes.

However, there has been at least one attempt to look for potential ETI signals in archival X-ray data[4]. As you probably guessed, no obvious ETI signal was found.

X-ray or Gamma-ray signals would have some advantages over radio or optical, but they come with their fair share of disadvantages too.

X-ray and Gamma-ray SETI has been called highly speculative. But that could also be said of radio and optical signals. After all, there is no way of knowing what technology ETIs will use until we find a signal.

There are no currently running or planned search programs in the X-ray or Gamma-ray parts of the spectrum. Financial resources are limited, and we have to choose where to look for signals. But it would be wise to keep in mind that there are still other parts of the electromagnetic spectrum beyond radio waves and visible light. It might be worth to take a look there in the future.

Technosignatures

Electromagnetic radiation is not the only place to look for signs of ETIs.

Mainstream SETI always did and still does favor searching for electromagnetic signals. And there is solid reasoning for this approach because, from the standpoint of current day technology, it seems easier to send an electromagnetic signal than to send probes or even ships to another star system. However, it is important to keep in mind that this is merely an assumption. There is still no way of knowing how an extraterrestrial civilization will behave and what they will be capable of.

Therefore, it cannot be ruled out that ETIs would send probes to our solar system to observe us, and maybe establish contact with us once we are able to detect them. Ronald N. Bracewell proposed such probes for communicating with other civilizations in 1960. They would contain a specific amount of information that the civilization which built the probe wants to share. These probes would be autonomous and would be controlled by an artificial intelligence. Such a probe would then wait and monitor the life forms of interest. Once they cross a predefined technological threshold, the probe would establish contact. They could also be stationed somewhere (like on the Moon or in the Asteroid Belt) so that only civilizations capable of finding and reaching them would get to know their information. Today, all such, still purely hypothetical, probes are called *Bracewell probes*.

You might think that we would know if there is ETI technology in our solar system. But in fact, we do not know with certainty that there is no extraterrestrial presence, be it in the form of life or technology, in our solar system. Especially if we define our solar system in a way that it includes the Oort Cloud (essentially a shell around the solar system made up of lots of comets), which has not even been directly observed so far.

But there is a field of research that deals with precisely these kinds of things. It is called SETA (Search for Extraterrestrial Artifacts). The name was coined by Robert Freitas and Francisco Valdes in the 1980s. They argued for searching our solar system for signs of extraterrestrial technology, and proposed specific spots and orbits to do so. They specifically recommended searching at Lagrange points of the Earth-Moon and Earth-Sun system (these points are essentially positions where a probe could stay in a very stable orbit for long periods of time without the need to continually correct its position, which would require energy).

There have been searches in the past looking at these and other spots. All have turned out empty so far. But we must not forget that our solar system is big, and potential probes might be tiny. In fact, the rise of nanotechnology here on Earth makes it seem likely that extraterrestrial probes might be very small, maybe just a few centimeters in size or even smaller. Even if there were a few of them in the vicinity of the Earth, there's no way we would have found such small objects. But even if the probes were bigger, if there would be a probe a few meters in diameter and dark on the outside somewhere near us, maybe hidden in the asteroid belt between Mars and Jupiter, it would still be damn hard to find that thing.

The search for probes or any other form of alien technology or signs of past activity specifically in our solar system has recently been called SETV (Search for Extraterrestrial Visitation). The term SETA, on the other hand, now includes looking for any sign of extraterrestrial technology, in our solar system and beyond.

It includes searching for alien probes that might be in our solar system to observe us or establish contact once a specific event happens (e. g., we pass a certain technological or mental threshold). Furthermore, it includes looking for monuments and active or abandoned bases on planetary bodies, moons or asteroids in the solar system. It does also include looking for signs of past technological activity like abandoned factory complexes or debris from mining activities.

Outside of our solar system, it includes looking for signs of astroengineering, which describes large-scale engineering projects that would be observable on astronomical scales. The most famous example of this is a *Dyson sphere*, named after physicist Freeman Dyson who first proposed the concept in 1960. It is essentially a shell around a star to harvest all or most of its energy output. It is assumed that the radiation from a *Dyson sphere* would look rather unusual compared to what you would expect to see from a regular star. Several searches for these kinds of odd radiation have been conducted. But recent advancements in materials science may lead us to question if we are actually looking for the right type of radiation. It now seems possible that advanced ETIs can create materials that will behave differently than we expect, and don't emit the kind of radiation we are looking for.

Another form of megastructure to look for is a *Shkadov thruster*. In simplified terms, it is a gigantic light sail that is used to move the whole star (very slowly).

Both the *Shkadov thruster* and the *Dyson sphere* are types of so-called *stellar engines*, with the *Shkadov thruster* being a Class A *stellar engine* and the *Dyson sphere* being a Class B *stellar engine*. There is also a Class C, which is essentially a combination of both.

The first attempt at SETA dates back even many years before the concept was proposed by Freitas and Valdes and even before Bracewell's proposal. Percival Lowell's effort at the end of the 19th century to study a civilization on Mars based on signs of their technology, specifically canals of supposed artificial origin, was in principle a SETA effort. Due to better telescopes and high-resolution images taken from spacecraft in orbit around Mars, we now know that there are no artificial "canals" on Mars. What Lowell (and others) saw was merely an optical illusion. Remember, this was at the end of the 19th century. Not only was the resolution of the telescopes not that great compared to today's instruments. Also, observations of Mars were not made with the aid of photography but by eye. Astronomers had to stare through the telescope for long periods of time.

SETA and SETV actually belong to the field of SETI, as they aim to find evidence of ETIs via a scientific approach. They are just looking for other signs of ETI technology. Rather than looking for radio transmissions or laser pulses, they are looking for actual technology within our solar system or outside of it.

However, they were never really an accepted part of mainstream SETI. This is unfortunate because the field has been at best neglected and at worst ridiculed, although it is based on valid assumptions on why artifacts could exist and where to look for them. But some influential members of the SETI community have argued that electromagnetic radiation is clearly superior to sending probes, and therefore no civilization would attempt to send probes to establish contact with us.

Although the arguments of the "electromagnetic faction" are plausible, there are also compelling arguments that make the search for artifacts credible. According to Freitas, the use of probes is even superior to the use of electromagnetic waves for various reasons[5].

Some of the reasons for using probes instead of (or maybe in addition to) electromagnetic waves brought forth by Freitas include the information gain of the sending civilization. In the case of a beacon sending out electromagnetic signals, the sending civilization would only get information in the form of an answer from another civilization. But this necessitates that a planet receiving the signal not only has life but intelligent life. Also, it has to have passed a certain technological threshold of possessing the capability of recognizing and answering such a signal. Furthermore, the receiving civilization must not only be capable of but also be willing to answer. With a beacon, the sending civilization would learn nothing about life forms that don't fulfill these criteria.

If a probe has a high level of artificial intelligence and self-repair or even self-replicating ability, which seems even more probable today, the sending civilization would have to put in the energy and raw materials to build such a probe. But once it is sent off into space towards its (initial) target, they don't have to put any more energy or raw materials into it. If the probe needed more power or resources than its initial load, it would gather these resources on its way. You don't have this advantage with a beacon that sends out electromagnetic signals. A beacon needs to be powered and maintained during its whole lifetime.

The interaction between a probe (as a messenger of the sending civilization) and the receiving civilization would be way faster due to the proximity of sender and receiver. We could send a signal to a probe within our solar system and receive an answer in a matter of seconds or minutes, compared to decades or millennia when transmitting a signal to another star system. The artificial intelligence of a *Bracewell probe* could control which information it gives us at what stage so that we are not overwhelmed by the information we receive. Of course, an ETI that is transmitting a radio signal could also choose which data to send based on the answers it gets. But those answers take a much longer time to another star than to a place within our solar system.

Also, if they would send a signal that is not very sophisticated (like a beacon solely to get attention), it might not be entirely clear if it is indeed an ETI signal. Take FRBs (Fast Radio Bursts) as an example. These could be (and probably are) of natural origin. But they might also be a ping of the galactic Internet. Granted, this seems unlikely. But it is just used as an example to show that an ETI signal might not be understood as what is actually is. If, on the other hand, we would find a piece of technology out there that did not launch from Earth (and we didn't put that much out there so far, especially in the outer solar system) there would be no question about whether this is proof of ETI or not.

But despite these advantages, SETA is not a welcome guest on the SETI party. And there is another reason for that. For a lot of scientists, SETA and especially SETV sound similar to Ufology. And that is a topic that has a terrible standing with scientists, and for good reasons.

The term UFO means Unknown Flying Object. So, any object in the sky that is observed and not immediately identified is per definition at first a UFO. Ufology then is, according to Wikipedia[6]

"... the study of reports, visual records, physical evidence, and other phenomena related to unidentified flying objects (UFO)."

In principle, this sounds like a legitimate thing to do. And it is. Still, Ufology is not considered to be a real science by most people and is outright rejected by almost all scientists and considered a pseudoscience.

The main problems of the field are that a lot, if not most, of the members of the UFO community equate UFO with alien spacecraft and that the evidence brought forth to support such claims is never really compelling and often fraud.

There is nothing wrong with seeing an Unknown Flying Object in the sky and wanting to find out what it is. But to many members of the UFO community, a UFO is considered an alien spacecraft by default. If you see something and don't know what it is, how can you immediately conclude that it is a spaceship from another civilization?

"Extraordinary claims require extraordinary evidence"[7]. But the evidence is not at all extraordinary. There is no compelling evidence for the claim that even one of the observed UFOs was an alien spacecraft. Note that witness testimony alone, especially from people who are not trained in any relevant field, is usually not considered tangible evidence. Even worse, a lot of the presented evidence is, after scientific scrutiny, easily disguised as fraud.

It is entirely possible that spacecraft or probes from another civilization could visit us here on Earth. But why would they hide somewhat but not really? If they have the capability for interstellar travel, they would have the technology to hide from our eyes if they wanted to. If they wanted to be seen, they would send a signal or land on Times Square and not somewhere in the countryside and cut circles in some cornfield. And if they didn't want to be seen, we wouldn't see them.

Another point is that the imaging capacity of the world has expanded several orders of magnitude in the last decades, especially with the spread of smartphones. Therefore, you would expect to see a drastic increase in the amount (and quality) of images and videos of UFOs. But that is not the case. There are still not significantly more recorded UFO sightings than a few decades ago.

These reasons have led to the almost universal refusal of Ufology by the scientific world. Although at its core it could be a viable field of research. But for the above reasons, it is hard for the scientific community to take Ufology seriously and it is hard for legitimate scientists to pursue real scientific research in the field of Ufology when the whole field is met with rejection or even ridicule by almost all other scientists.

The so-called *paleocontact hypothesis*, which states that Earth has been visited (and influenced) by ETI in the past is plagued with similar problems. It is just as plausible that the Earth was visited by ETI a few thousand years ago as it is that the Earth is being visited now. But this hypothesis attracts a lot of fraud, and potential evidence is not compelling. Every possible clue presented in support of this hypothesis can be explained in another way that doesn't involve ETI.

In the recent years, probably due to the many found exoplanets, SETI is becoming more and more accepted in science and government circles. But that hasn't been the case in the past decades. For a long time, the legitimacy of SETI as a scientific endeavor has been seriously questioned again and again. SETI scientists regularly had to explain that what they were doing had nothing to do with UFO sightings. No wonder they didn't want something that sounded so much like UFOs to be accepted as mainstream SETI. I mean, come on, Search for Extraterrestrial Visitation sounds a bit like backyard abduction. I think the term Search for Extraterrestrial Artifacts is the better of the two.

But fortunately, just as SETI has become more mainstream in recent years, the acceptance of SETA is also slowly rising, especially in Europe, where discussions about the Search for Extraterrestrial Artifacts are happening at reputable institutions like *ESA* or the German *Research Network Extraterrestrial Intelligence*.

Of course, this way of communicating is not perfect either. An ETI would first have to build such a probe capable of interstellar travel. They also would have to be capable of programming a reliable artificial intelligence to control the probe. Then it has to get to its target location without getting damaged. And that voyage would take some time as the probe, unlike electromagnetic waves, would probably not travel at a speed anywhere near the speed of light. And if all goes well and the probe arrives at its destination and establishes contact, it would still be limited by its original programming, and the amount of information it got from home. So we couldn't actually talk to the civilization which sent the probe. After all, the whole point of sending such a probe is to shorten the distance between the sender and receiver of a signal, so we wouldn't have to wait thousands of years for an answer.

That is unless, of course, such a probe would have the technology to communicate with faster than light signals. In this case, its primary purpose would probably be to teach sufficiently evolved civilizations how the adults in the galaxy talk to each other and spare technologically less advanced civilizations decades of looking for these, from their perspective prehistoric, radio waves.

The possibility of interstellar probes seems even more likely now than it was when it was first proposed by Bracewell in 1960. With the technological advancements since the 1960s, it seems not that far off that we humans could be able to send our own *Bracewell probes* to other stars somewhere in the coming decades. If they will be able to travel faster than light remains to be seen. At least our current understanding of physics makes this seem extremely unlikely. But the point is, you don't need faster than light travel capability to send out a *Bracewell probe*.

Neutrinos

For some time now, researchers have thought of another way for ETIs to send signals. And that is via neutrinos. A neutrino is an elementary particle with very little mass that is electrically neutral. It interacts with almost nothing, which means it can pass through huge amounts of matter and continue on its way unimpeded. This makes it very hard to detect but also an interesting possibility for interstellar communication.

In 2008 John Learned together with Sandip Pakvasa and A. Zee proposed a way that a civilization could communicate with neutrinos[8]. They were not the first scientists to seriously think about the possibility of neutrinos as a means of interstellar communication (that was M. Subotowicz in 1979). But what is especially interesting here is that not only should we be able to create such a signal in the foreseeable future, but also that our currently existing neutrino detectors should be able to detect them.

In an experiment conducted at the US-based *Fermilab* in 2012, scientists actually transmitted information via neutrinos[9]. Although this experiment required tons (literally) of sophisticated equipment and the rate at which information was transmitted was really really low, it does prove that communication via a neutrino beam is possible, not just in theory but practice.

So while communicating with neutrinos is more complex than using radio waves, it is a possibility. And if today's detectors could detect such a signal (at least a signal like the one proposed by Learned, Pakvasa, and Zee) why not keep an eye out for them? Especially because you don't have to build specialized SETI devices or buy expensive observing time on neutrino detectors. All the data on incoming neutrinos is recorded at these detectors. You would just have to analyze the data and look for something that looks like an ETI signal.

Gravitational waves

Gravitational waves are waves in the fabric of spacetime (the three dimensions of space, that you're familiar with, plus the dimension of time together are called spacetime). So, rather than moving through space, as electromagnetic radiation does, they wobble spacetime itself. Such waves are created by the accelerated motion of masses. In fact, my fingers flying over the keyboard right now are creating gravitational waves. But they are way too faint to be measured any time soon.

What we can measure, however, are gravitational waves created by astronomical phenomena. Gravitational waves were first predicted about a hundred years ago by Henri Poincaré. They also are a result of Einstein's General Theory of Relativity, which was published a few years later.

But the actual detection did not happen until 2015. Scientists used an incredibly sensitive instrument (the *Laser Interferometer Gravitational-Wave Observatory*, or *LIGO*). And still, it needed black holes to collide to produce gravitational waves strong enough to be detected. And these are really heavy objects. That might give you a sense of what masses are required to create gravitational waves which can be measured with today's technology.

Is it possible that ETI might communicate via gravitational waves? Yes, of course. They have their advantages. Mainly, that they pretty much don't care about anything in their way. They move through dust clouds, planets and everything else in their way almost completely unhindered. That makes them interesting for long-range communication. But the downside (and that's a big downside) is the masses required to create gravitational waves that are strong enough to be received by someone. Like slamming together black holes or neutron stars. The energy needed to do that regularly is unbelievable.

While it is not unthinkable that a really advanced civilization could be capable of doing that, it doesn't seem very practical. In particular when you have so many other options to communicate that require tremendously less energy. Especially because with gravitational waves, you are still limited by the speed of light.

We should be looking for gravitational waves. They can tell us a lot about the universe that other methods cannot. And while it is unlikely that ETIs would communicate with gravitational waves, it doesn't hurt to search the data for possible signals. Luckily, as with neutrinos, you don't have to buy expensive observing time for SETI, you can just analyze the data that is recorded.

Other means of communicating

There are still other possible ways for ETIs to communicate.

They might use quantum entanglement. This is a phenomenon well known to us, but not really understood. For communication between ETIs, it would have a considerable advantage. In principle, it should be possible to send messages instantaneously from sender to receiver, regardless of the distance between them. Now, this is a massive advantage over all the other techniques we discussed. Also, no one except the sender and the intended receiver would be able to see the message. Or so it seems from our current understanding of quantum entanglement. But that does mean that ETIs deliberately trying to contact civilizations that are not aware of their presence wouldn't use technology like that. Maybe advanced ETIs would use this method to communicate among themselves. But they would use a different medium of transmission to contact civilizations like our own.

Even beyond that is the possibility of communication via Zeta rays. Okay, Zeta rays are not really a thing. It is a term used to describe physics that lay outside of the Standard Model, which is the theory we use to explain how the basic building blocks of matter interact. The term is usually not used in a negative way, but to describe all physics that might be possible but are not part of the Standard Model we use today. After all, physicists at the Large Hadron Collider (a giant particle accelerator located at *CERN*, the *European Organization for Nuclear Research*) are actively looking for evidence of new particles that we do not know about yet.

We know a lot about the universe, and we're making progress every day, but there are still a lot of holes in our theories. This leaves a lot of room for possible ways of communication that we have absolutely no idea about. We could receive such signals all day long, and we wouldn't be able to recognize them as a signal at all. But science is moving forward, and with it, our body of knowledge gets bigger and better every day. If there are other ways to communicate, I am optimistic that we will eventually figure them out. Until then, we have to look at the places we already know about.

Signs of life in general

The search for extraterrestrial intelligence is a fascinating endeavor, and finding signals from another technological civilization would be a wonderful thing. But extraterrestrial life doesn't necessarily need to be intelligent. Even finding signs of microbial or plant-like life elsewhere in the cosmos would be fantastic.

Exoplanets

The interest in SETI has increased in the last years because of significant advancements in another field, namely the search for extrasolar planets or exoplanets. This is the search for planets around other stars.

There have been some "rogue planets" found that don't revolve around a star. These are also called exoplanets. So, exoplanets are actually all planets that are outside of our solar system. But most of them seem to be around other stars and not wandering alone.

The search for exoplanets is an exciting and fascinating area of study. And as one of the primary goals of this field is to find signs of life on other planets, it has to be included in this book.

In the 20th century, scientists believed that our solar system was an exception and that there were no (or very few) planets around other stars. That all began to change in the 1990s with the confirmed detection of an exoplanet (51 Pegasi b). There actually was a detection in 1988, but it was not accepted as being an exoplanet at that time. From that time on, research efforts drastically increased and so did the number of known exoplanets.

As of September 2018, *NASA's Exoplanet Archive* lists 3,791 confirmed exoplanets. Most of them were discovered by *NASA's Kepler* mission. And there are still a lot more that await confirmation. So the number is expected to rise.

We have found so many planets around other stars (and even a lot of systems with multiple planets) that it seems clear by now that our solar system is not the exception. Rather, it is assumed that almost every star in our galaxy has at least one planet.

There is a pretty big uncertainty about the number of stars in our galaxy because we can't just look in a telescope and count all of them. Some stars are very bright, others are dim. There are massive concentrations of stars where it is difficult to see the single stars because they are so close together. There are nebulae and dust clouds that obscure our view, and not to forget the massive black hole in the middle of the galaxy that makes it pretty hard to see what's behind it, among other things. So the number of stars in our galaxy is estimated based on statistics. The current estimate is 250 billion +- 150 billion.

As you can see, there are a lot of stars in our Milky Way alone. And if almost all of these stars have one or more planets, that leaves us with a lot of planets and therefore potential places where life might have arisen. A lot of these planets do not look like Earth. But there are still enough worlds where *life as we know it* should be able to survive. Now that we know that our solar system is not an exception and there are so many planets out there, naturally this increased interest in SETI efforts.

There are several methods to detect exoplanets. It is possible to image them directly. In other words, it's possible to take a picture on which you can see the planet beside its star. But that is pretty difficult with our current telescopes. Planets are not just small compared to their stars, the stars are also a whole lot brighter. That makes them really hard to spot. The methods with which most of the exoplanets we currently know of were detected are the transit method and the radial velocity method.

When a planet moves in front of its star as seen from Earth, we can measure a dip in the brightness of the star. The change in the star's light is around 1/10,000. This tells us not only that there is a planet around that star but even the planet's size. The larger the planet, the greater the dip in brightness. This is called the transit method.

The Earth moves around the sun with about 30 km per second (67,000 mph). The reason why Earth does not just fly out of the solar system and into interstellar space is gravity. In this case, the gravity of our sun. It pulls on the Earth and thereby forces it into a nearly circular orbit. But gravity works both ways. The sun doesn't just pull on Earth, Earth also pulls on the sun. The power of this gravitational pull is determined by the mass of the object. Therefore, the sun draws a lot harder on Earth than the Earth does on the sun. This effect can be beneficial in finding exoplanets. Exoplanets too pull on their host stars. This causes the star to wobble a little bit. It's a tiny bit, but as it turns out, it is enough for us to measure it. If we see a star wobble, we can infer that there is a planet around that star. And we can even get the mass of that planet. This is called the radial velocity method.

There are other ways of finding exoplanets, but the transit method and the radial velocity method are the ones with which we discovered the overwhelming majority of exoplanets we currently know of.

You can learn more about these and the other methods to detect exoplanets at a fantastic *NASA* website called *"5 ways to find a planet"*[10].

Habitable zone

It is exciting and useful to know about all these exoplanets in their own right. But of course, one crucial aspect of the search for exoplanets and especially important for this book is the question "Is there life on any of these planets?"

To find out if there is life, we first have to determine if there could be life. In other words, before we ask if the planet is inhabited, we have to find out if the planet is at all habitable. The topic of planetary habitability has been intensely studied over the course of the last years. In the beginning, we thought we pretty much need a clone of Earth around a Sun-like star to have any chance of finding life. Our current understanding allows *life as we know it* on a lot more planets around different types of stars. Let alone the possibility of *life as we don't know it*.

With all the recent announcements of exoplanet findings, you have probably heard of the concept of a habitable zone. That is the zone around a star where it is not too hot and not too cold for liquid water to exist. Pretty much everywhere on Earth we looked, we found life. We can see a lot of different life forms on Earth thriving in environments that were just one or two decades ago thought unimaginable to harbor life. Yet we have found it. But what was common among all forms of life we found was that they all required liquid water. Some a lot, some just a tiny bit, but they all needed water. And that water needs to be in liquid form. So, it is reasonable to assume that life on other planets would also need liquid water.

Of course, all life forms we know of are from Earth. They all evolved here and adapted to this environment. So there is the possibility of life that is based on entirely different requirements. But we have to start somewhere.

Where the habitable zone actually is around each given star depends on different factors.

It depends on the type of star. For smaller, cooler stars the habitable zone is closer to the star than for large, very bright ones. Not every star is suitable, however. There are things like neutron stars. These are dead stars, or star corpses if you will. They are small but have a very high mass. Some of them emit enormous amounts of radiation. Although there have been planets found around these types of stars, nobody really hopes to find life on these planets.

It also depends on the atmosphere of the planet (think greenhouse effect). Venus in our solar system, for instance, is very hot due to a greenhouse effect. Yes, it is a bit closer to the sun than the Earth is, but that is not enough to explain the temperature difference. The surface temperature on a typical Venus day is around 864 degrees Fahrenheit (462 degrees Celsius). That is hot! As a side note: Discovering this runaway greenhouse effect on Venus was what brought us to the idea that we might be causing our climate here on Earth to change due to the stuff we blow in the air. So, a planet with an enormous greenhouse effect could sustain liquid water farther out from its host star. That makes the habitable zone around a star bigger.

And that's not all. The habitability of a planet can also be influenced by an internal heat source. That heat can come from different processes. It could be leftover heat from the formation of the planet (or moon). It could be radioactive decay, or it could be something like tidal flexing. That last one is an interesting way to produce heat, and it can be observed within our solar system. Let's take a look at the moons of Jupiter. We'll pick Europa as an example because it is a moon where we actually might find life. This moon is heated by Jupiter's enormous gravity. As Europa revolves around Jupiter, it is stretched and squeezed by Jupiter's gravity. This produces heat. And it might just be enough heat to allow a substantial underground ocean on Europa, which would make the moon potentially habitable. Note that, contrary to popular belief, direct sunlight is not necessarily a requirement for life. On Earth, there are life forms that live on the bottom of the Ocean and never see any light from the sun. This makes the habitable zone around a star even bigger.

However, the regions where planets need an internal heat source and therefore probably don't have liquid water on their surface but further down, are not universally considered to be part of the habitable zone. Nevertheless, such planets (or moons) could be habitable.

As you can see the area around a star where planets might be habitable is pretty big. But of course, most exoplanet hunters are hoping to find a planet with very similar conditions to that on Earth. We can (and should) speculate about all the kinds of environments where life could thrive. But so far, we have only one single planet where we have proof of life, and that is the Earth. Therefore, if we were to find a world with conditions similar to that on Earth, we can say that on this planet there could be life. It remains to be seen if there is life, but at least we know for sure that the conditions would, in principle, allow for life to evolve, because it has happened at least once before, here on Earth.

Okay, we know that for a good chance to find life on another planet we should look for planets that are in the habitable zone of a star. We have looked at how we define the habitable zone for any given star. But how do we find out if a planet orbits within these limits? It is not that hard to calculate some characteristics of a planet once you have good observational data. Getting this data is a challenge, but we're getting better at it.

Kepler's laws (specifically the third one) allows us to calculate the size of the planet's orbit if we know two things: how long it takes the planet to orbit its star once, and the mass of the star. You can get the mass of the star by measuring its luminosity and temperature (if you want to know more about that take a look at a *Hertzsprung-Russell-diagram*). The time it takes the planet to orbit its star once is easy to determine. We just look at where it was when we first saw it and wait until it is in that exact spot again. For instance, if we found the planet through the transit method, we can see a dip in brightness in the light curve of the star. Then the dip disappears as the planet moves out of our line of sight. When the planet comes back around it will produce the same dip in brightness again. The time between the two dips is the time it takes the planet to orbit its star once (also known as its period). And with that, we can calculate the size of its orbit. What that tells us is the distance of the planet from its star, which is very important for determining if the planet is in the habitable zone.

When we know the temperature of the star and the size of the orbit of the planet we can calculate the temperature of the planet. Of course, the temperature can be significantly influenced by the planet's atmosphere, but this calculation sets some upper and lower limits on the temperature of the planet.

The size of the planet can easily be determined using the transit method. The larger the world, the deeper the dip in brightness of the star when the planet moves in front of it.

The mass of the planet can be determined using the radial velocity method. The larger the pull of the planet on its host star, the larger the mass of the planet has to be.

The density of the planet can be calculated by simply dividing the mass by the volume. Knowing the density lets us distinguish between gas giants and rocky planets for instance. Rocky planets are currently more interesting to us as we are looking for planets similar to Earth.

These values give us a reasonable estimation of the planet's habitability. We can assess if the planet is, in principle, habitable. If the planet is really habitable depends on these factors but some others as well (like the atmosphere). If the planet is actually inhabited is a lot harder to tell.

But even if the inhabitants of a planet are not transmitting microwave signals or laser pulses our way (either because they don't want to or they can't), there are other ways to find out if there is life on a planet. They are not as definitive as a radio signal telling us "we are here, wanna talk?", however.

Analyzing exoplanet atmospheres

Now that we know of all these exoplanets (and will very likely find a whole lot more in the years to come) we have another possibility to search for life in general, not just intelligent life. And that is by analyzing the atmospheres of exoplanets.

We cannot image most of the planets directly (yet) because their host stars usually shine millions of times brighter than the planet. But there is another way that can tell us more about the atmospheres of these planets. For that, we use a technique called spectroscopy.

We already know that all electromagnetic waves are essentially the same, the difference being their wavelength (or frequency or energy; they are all tightly linked together and are merely different ways of saying the same thing). In the visible part of the spectrum, different colors mean different wavelengths. It turns out that different molecules (like methane, oxygen and so on) absorb light at characteristic wavelengths. Let's say you shine light through a cloud of oxygen. Then you can see on the other side of the cloud that there are certain parts of the spectrum missing that initially went in on the other side. Turns out these missing parts in the spectrum are always the same for a given molecule. This, in turn, means that you can infer through which gases the light went by looking at what parts of the spectrum are missing. And that is precisely what is used to see what gases are in the atmospheres of exoplanets.

As you already know, there are several ways of detecting planets around other stars. Of these methods, it is the transit method that is most interesting for analyzing the atmosphere of an exoplanet. When a planet orbits its star directly in our line of sight, which means the planet moves directly in front of its host star as seen from Earth, it dims the light that we see of that star for the time that it is in front of it. But if the planet has an atmosphere, then some of that light (of the star) will pass through the atmosphere of the planet. This means, when we detect the light from the star, we do not only see a dip in brightness. We can also see these absorption lines characteristic for some molecules.

Stars do have these absorption lines too. So how do we know which absorption lines are from the star (i. e. molecules that are present in the star) and which ones are from the planet (i. e. molecules that are present in the atmosphere of the planet)? That's simple, we just wait a bit. When the orbit of the planet is so that it moves in front of the star as seen from our point of view, then there has to also be a time when the planet passes behind the star. When we compare these spectra (one with the planet and one without it, when it is behind the star and therefore not visible to us), we can see the difference. From that, we can infer which molecules are present in the star and which ones must be in the atmosphere of the planet.

Our current (and planned) telescopes are not nearly powerful enough to snap a photo of an exoplanet on which you would be able to see things like cities. But what we can do is analyze the atmospheres of some exoplanets. So we have to find a way to find out if life is present on an exoplanet by looking at its atmosphere. Or, more accurately, by looking at the most abundant molecules in the atmosphere, because gases that are only present in tiny amounts are pretty much impossible for us to detect.

So, how do we answer the question "is there life on this planet?", just by looking at the atmosphere? That's where the concept of biosignatures comes in. According to Wikipedia[11]

"A biosignature ... is any substance – such as an element, isotope, molecule, or phenomenon – that provides scientific evidence of past or present life."

Some gases might tell us about the presence of life. These biosignature gases are defined as a gas that is produced by life. It also has to accumulate in the atmosphere to a level that we can detect (remember that tiny amounts of a gas are not recognizable to us).

There are certain gases or combinations of gases that we think might be a good indicator of biological activity on a planet. Note that we are not necessarily talking about intelligent or even technological life here. Of course, it would be amazing to find another intelligent, technological civilization out there. But finding microbial life or a planet full of some form of animal would also be a pretty big thing.

Suspicious are gases that are not in chemical equilibrium. This means, some gases shouldn't be there (for long) unless they are refilled continuously by something that produces them. That something might be biological or non-biological. Oxygen, for example, is known to be produced in large quantities by plants on Earth. But there are also geological and atmospheric processes that produce oxygen. So, while oxygen can be indicative of life, it is by no means definitive proof.

A lot of gases have been suggested to be valid biosignatures. Among them are oxygen, ozone, and methane. We must, however, be very cautious when we detect one or more of these gases in the atmosphere of an exoplanet. There are known non-biological processes that produce at least some of the proposed biosignature gases. Also, there can be processes that we don't know of yet, which create these gases but have nothing to do with life. It will surely be exciting to find some of these biosignature gases in the atmosphere of an exoplanet. But we must examine the data carefully and do multiple thorough observations of that planet before we even seriously consider that it could be inhabited.

Finding life from afar by looking for biosignatures is hard. Luckily, after we began finding ever more exoplanets, this field of research got a massive boost. There are a lot of scientists thinking long and hard about how to find life by looking at the atmospheres of exoplanets. Let's hope they can put their theories to the test very soon.

> **Kepler**
>
> Most of the planets we know of today were discovered by *NASA's Kepler* mission. *Kepler* is a space telescope with a constant field of view. It continuously looked at about 150,000 stars. It monitored these stars for dips in brightness produced by a planet passing between the star and the telescope (the transit method). The spacecraft was launched in 2009 and operated until 2013 when the mission had to be ended after the failure of 2 of 4 reaction wheels. Without 2 of its reaction wheels, it was no longer able to fulfill its original science goals because it could not be stabilized enough to continue pointing at its field of view with enough precision. That meant the end of the *Kepler* mission. *NASA* repurposed the telescope in 2014, and it now is on its *K2* mission with a different mode of operation and different science goals. The original *Kepler* mission may have ended, but it was a hugely successful mission that found most of the planets we know of today. Indeed, it has found 2,656 of the 3,791 confirmed exoplanets, plus lots of candidate exoplanets that still await confirmation (during its original mission and the *K2* mission; as of September 2018). Unfortunately, we won't be able to use this telescope much longer as it is expected to run out of fuel in the second half of 2018.

TESS

There is another space telescope besides *Kepler*, specifically designed to find exoplanets that has already begun operations. It is called *TESS*, which stands for *Transiting Exoplanet Survey Satellite*. As the name suggests, this telescope also uses the transit method to detect exoplanets. *TESS* launched in April of 2018 and officially began science operations on July 25, 2018. It will spend two years looking at 200,000 of the brightest stars in our galactic neighborhood. *TESS* is expected to make thousands of new exoplanet discoveries. Currently, it is estimated that about 300 of these planets will be no more than twice the size of Earth and therefore of particular interest for the search for extraterrestrial life.

JWST

There is a telescope that astronomers have been waiting for for years. And they will have to be patient a bit longer because it is still not launched. It is supposed to be the successor of the famous *Hubble Space Telescope*, and it's called the *James Webb Space Telescope* (*JWST*). Although it is called *Hubble's* successor, it is an infrared telescope (contrary to the *Hubble*, which operates mainly in the visible part of the spectrum, plus a bit of near-infrared and ultraviolet). The construction of the telescope has already taken several years longer than initially planned and costs have exploded. But still, when it finally launches, it will be the most powerful telescope at our disposal. It will have a 6.5-meter primary mirror, which is enormous for a space telescope. For comparison: the Hubble Space Telescope has a 2.4-meter mirror.

> The *JWST* will, among other things, look for exoplanets. It will be able to find planets around other stars using the transit method. It will also be powerful enough to analyze the atmospheres of these planets. It is indeed so powerful that *NASA* expects it will be able to take direct images of some exoplanets. Even with the *JWST*, we will not be able to see a lot of surface features on an exoplanet. Rather, it will just be a dot. But that dot might tell us something about seasons, weather and maybe even vegetation on that planet. All of these have to be large-scale, of course. We won't be able to take a look at the flowers in some alien's front yard. It will probably still take a while before we see the first images from the *JWST*, however. After being postponed again and again, the launch is currently planned for 2021.

Our solar system

We already looked at a host of different methods and techniques to find life somewhere else in the universe. But we might not have to look that far out, after all. While there is apparently no humanity 2.0 somewhere in the solar system, we already found out that we cannot say with certainty that there is no extraterrestrial technology here.

But we shouldn't just be looking for signs of technology that some ETI from another star system sent here. What about life that evolved right here, in our very own solar system?

We'll now take a look at some interesting places where life might have formed in our solar system. This list is by no means complete. There are other places where life could have developed, but I've picked the ones that seem most promising right now.

Mars

Mars. Of course. We've been wondering about life on Mars for what feels like forever. Remember Percival Lowell and his Martian Canals?

To this day, we have found no evidence of past or present life on Mars. What we have discovered are conditions that would make Mars, in principle, habitable. At least in the past.

Since we've become a civilization capable of (rudimentary) spaceflight, we've sent many probes to Mars:

- *Mariner 4* (1964)
- *Mariner 6 & 7* (1969)
- *Mariner 9* (1971)
- *Viking 1 & 2* (1975)
- *Pathfinder* (1996)
- *Mars Global Surveyor* (1996)
- *Mars Odyssey* (2001)
- *Spirit & Opportunity* (2003)
- *Mars Express* (2003)
- *Mars Reconnaissance Orbiter* (2005)
- *Phoenix Scout* (2007)
- *Curiosity* (2011)
- *Maven* (2013)
- *Exomars* (2016)

And these are only the successful missions. You can find an excellent overview of missions to Mars on *NASA's* Mars website[1]. And more missions are planned. If there is or were life on Mars, we'll eventually find it.

Past and current missions have told us a lot about Mars. For instance, that it has underground water ice. At some spots, it is even at the surface.

Water in its liquid form might also be present on the surface today, but only for short periods of time. Any liquid water getting to the surface would freeze or evaporate almost immediately, due to Mars' very thin atmosphere (less than 1 % of Earth's air pressure).

Also, this thin atmosphere, along with a missing magnetic field, make it pretty uncomfortable for life on the surface, because there is no protection from cosmic and solar radiation.

But Mars is thought to have been warmer and wetter in the past, conditions more favorable to life, at least microbial life. Findings from *NASA's Curiosity* rover recently revealed that Mars would have been capable of supporting such microbial life billions of years ago. No one knows if it did, but it could have if such life really evolved.

Most scientists consider it very unlikely that there is life on the Martian surface today. However, it is very likely that Mars was once habitable for *life as we know it*. If it did evolve back then, we will hopefully find fossils of that life somewhere on Mars. And who knows, maybe Martian life has survived to this day and keeps on thriving below the Martian surface.

It is possible that we will find evidence of past or present life on Mars in the future if we look hard enough. Luckily, interest in Mars has stayed high over the decades, and there is no shortage of missions to the Red Planet.

Europa

Now we journey farther away from the sun, to Jupiter. But it is not so much Jupiter itself that's interesting for the search for life. Rather, it is one of its many moons. Most of its moons are pretty small. The four biggest ones are called Galilean moons because it was Galileo

Galilei who discovered them. One of these moons is called Europa.

This moon was not at all on our list of potentially habitable worlds not so long ago. Only when *NASA's Galileo* mission (there's that name again) discovered evidence of an ocean underneath Europa's ice crust in the 1990s did it become a promising target in the search for life.

Europa's surface is a thick layer of ice, water ice that is. The moon is massaged by Jupiter, which means that the interaction with Jupiter's strong gravity field leads to the squeezing and stretching of Europa. The effect is not so drastic, you cannot really see Europa getting squeezed with an average amateur telescope. But all this squeezing and stretching leads to the production of heat.

It seems like this internal heat is enough to keep water liquid under Europa's ice crust. Measurements of its magnetic field support this idea, because the data looks like there is a massive ocean of salty, but liquid water underneath an ice crust several kilometers thick. The exact dimensions are not known, but it seems like there probably is more liquid water on Europa than in the Earth's oceans. That's a lot of water. Images of Europa show cracks in the ice that also make the idea of an underground ocean plausible.

If there is an ocean, however, it is always under the ice. Several kilometers of ice. This means that no sunlight ever gets down there. Photosynthesis, therefore, is not an option for possible life on Europa. But we know that direct sunlight is not a prerequisite for life. We have found life here on Earth at the bottom of the ocean where no sunlight ever gets to. Europa might have similar forms of life. At least it is entirely possible that there could be life on Europa, deep down in its vast ocean of liquid salt water.

This life, if it exists, will probably not be easy to find. It seems out of reach with our current technology to send a probe there, drill down through several kilometers of ice and take a look around.

Luckily, we might not have to do that. The cracks in Europa's ice crust could make our job a lot easier. We have discovered something that looks like plumes of water vapor coming from Europa's South Pole. If that turns out to be the case, it should be enough to fly through these plumes to analyze the composition of the underground water.

And there is a planned mission that is supposed to do just that, among other things. It's called the *Europa Clipper* and is currently set to launch in the 2020s.

ESA will also launch a probe to the Jupiter system in the 2020s. It's called *JUICE* for *JUpiter ICy moons Explorer*. Contrary to the *Europa Clipper* mission, *JUICE* will not primarily target Europa. It will observe Jupiter itself and three of the Galilean moons, including Europa.

While Europa is usually the star of the show when it comes to potential habitats for life in the Jupiter system, the other two targets of *JUICE* are also worth looking at. These are Ganymede and Callisto. These two moons are similar to Europa in that they also seem to have an internal heat source and lots of liquid water underneath an icy crust. We'll see what *JUICE* will tell us about them.

Enceladus

Let's move on a little bit further out. Now we'll take a look at the moons of Saturn.

Enceladus is an exciting target for the search for life in our solar system. It too is thought to have liquid water beneath an icy crust. It is believed to be heated by squeezing and stretching by the gravitational forces of Saturn and the other moons. The presence of liquid water alone makes it interesting. But what's more is that Enceladus has geysers that spew out water vapor and ice.

That means, here is another moon where we have the chance to analyze an underground reservoir of water, without having to drill through kilometers of ice, by flying through these plumes.

In fact, *NASA's Cassini* spacecraft already did just that. Although it was not explicitly designed for such a maneuver, it was still able to make an amazing discovery. In the water vapor and ice ejected from the geysers, it found organic molecules. And these were not just small organic molecules, but also complex ones comprising hundreds of atoms. Organic molecules don't necessarily mean life, but on Enceladus, we have liquid water, complex organic molecules, and hydrothermal activity. That definitely sounds like an interesting place to look for life.

Titan

Another fascinating place to look for life in the Saturn system is Titan. Now, Titan is a special one. Titan has interesting similarities with Earth.

It has lakes, and it rains on Titan. It is the only body in the solar system known to have lakes besides Earth. Also, it is the only moon in the solar system that has a significant atmosphere. This atmosphere is composed mainly of nitrogen (just like on Earth) and a bit of methane.

The big difference is that it is frigid on Titan. We're talking about -292°F (-180°C) cold. Therefore, as you might have guessed, there cannot be lakes of liquid water at this temperature. And in fact, there aren't. We wouldn't rate Titan very high on the probability scale for *life as we know it*. However, there might be *life as we don't know it*.

The lakes and the rain are made of ethane and methane (essentially liquid natural gas). On Earth, where it is a lot warmer, they are gases. On Titan, these two behave similarly to the water on Earth, forming lakes and clouds and falling as rain. As hydrocarbons like methane are a possible substitute for water, there might be life on Titan. Just not *life as we know it*.

Is there life on Titan? Nobody knows. On Earth, we have no spots where temperatures are (naturally) that cold. But, in principle, life could be possible in this environment. It might be different from life on Earth, but it's possible.

There is currently no spacecraft set to launch to explore Titan further. There is, however, a proposal for a quadcopter style probe called *Dragonfly*. *NASA* will decide in 2019 if this mission is actually going to happen. If so, it would start in the 2020s and arrive at Titan about ten years later.

Until then, all we can do is use our telescopes to analyze Titan further.

Other places

There are, of course, many more places in the solar system where we could search for life (planets, moons, asteroids, even comets).

It has, for example, been proposed that there could be life in the upper regions of Venus' atmosphere. On the surface of Venus the average temperature is about 860°F (460°C), but in the upper parts of its atmosphere temperatures are low enough to allow the existence of life.

It remains to be said that we don't necessarily need to look very far in our search for extraterrestrial life in the universe. We might find it right on our doorstep, in our solar system.

It's speculative, yes, but that is true for all efforts in the search for life in the universe. It is a possibility that we should keep investigating. Although it is extremely unlikely that we will find another technological civilization living on one of the planets or moons in our solar system, there might still be microbial life or even some form of fish on one of the icy moons, for instance.

And the search for life in our solar system has the significant advantage that we can visit all these potential habitats right now, at least with probes. We might, and hopefully will, be able to do that with other star systems too. But for now, all we can do is use our telescopes to look at them from afar. Actually sending a probe makes it way easier to determine if there really is life.

Conclusion on where we should search

We've seen that there are many options on where we could search for life. We've seen that there are different ways and different places to look for extraterrestrial life, from our solar system to the other end of our galaxy and even beyond. We've seen that there are ways to detect life. And that looking for intelligence actually means looking for technology as this is the only way we can be confident that what we found is intelligent because microbes will not build lasers or radio telescopes.

We've come a long way, but we do still have a lot to learn.

Finding organic molecules on planets or moons in our solar system doesn't mean there is life. We have to send more probes or even humans to explore more.

Finding gases in the atmospheres of exoplanets that could be produced by life doesn't mean there is life. We will have to find all sorts of explanations, and only if they are all proven wrong can we be confident that we found life. But we should definitely look for these potential biosignatures and then move on from there.

As for intelligent extraterrestrial life, all in all, it has to be clearly said that all of the assumptions and all of the reasoning on where to search in the electromagnetic spectrum (or even outside of it) are all guesswork. We simply cannot know by which means ETIs will communicate among themselves or with us before we find a signal.

As you can see, places where we could look are plenty. Not just is space pretty big; we also have to decide where on the electromagnetic spectrum we want to search. And what will an ETI signal look like? Or should we look somewhere completely different like neutrinos? Or can't we see signals at all, no matter how hard we look, because we don't even know about the physics behind their transmitting technology?

Some of us prefer radio waves, some optical lasers, some completely different phenomena like gravitational waves. They all might be right, some of them might be, or none. We just don't know. But we will never find out if we don't look. And we have to start somewhere. So, any effort to look for ETI signals is better than not doing anything.

We can debate all day long about which approach is the best and has the highest likelihood of finding an ETI signal. But unless we actually do something, actually search somewhere, we will never find out.

And then there's the possibility that we've already seen a signal from an ETI and misinterpreted it as a natural phenomenon. Of course, the discovery of another technological civilization out there would be such a big deal that we should be cautious when interpreting our data. Therefore, it is good to almost automatically assume a natural phenomenon when we first see an interesting signal.

But there might have already been an artificial signal among the ones we picked up. The only way to discover that is to keep investigating the potential sources of interesting signals and find proof that it really is a natural phenomenon. And one day a signal might turn out to be not of natural origin.

SETI and "regular" astronomy are not opponents. I consider SETI to be a legitimate part of the scientific field of astronomy. And SETI researchers and other astronomers (and of course astrophysicists and astrobiologists and any other related field) should work hand in hand, exploring any interesting signal from their different perspectives to find the truth. Whether that's a yet unknown natural phenomenon or a call from an ETI. Every signal lets us learn something new about the universe that we are a part of.

I hope this section has given you a bit of insight about the immense amount of places where signals from ETI could be found. And because of that you hopefully can acknowledge that even almost 60 years of not picking up any radio signal from ETI doesn't mean they are not out there. All of our search efforts were beneficial. We now know some places where no ETI signal is broadcast. But there is still much more work to do before we can conclude (in fact, before we can even seriously consider) that we are alone.

Should we send signals ourselves?

With our current day technology, sending messages out into space is not a problem. But should we do it?

METI

Observing the sky in the hope of picking up a signal from an extraterrestrial civilization depends on the assumption that there are civilizations that actively send out messages for others to pick up.

Although it is possible that we pick up signals which are not intentionally sent our way (think of our television signals, for instance), these signals would be way too weak for our current instruments to register.

So, the obvious question is: What if everybody is listening and nobody is sending? There could be lots of civilizations, and we wouldn't notice.

That's one of the reasons for METI efforts. METI stands for Messaging to Extraterrestrial Intelligence. It is pretty much what the name suggests. METI is about intentionally sending signals into space for other civilizations to pick up.

SETI is listening, METI is speaking.

However, deliberately sending messages or, more generally, making our presence known to possible ETIs is a very controversial subject.

The proponents of METI want to establish contact with other civilizations for the benefit of humanity and also to show possible ETIs that they are not alone.

It is often assumed that other civilizations will be more advanced than us. But we might just as well be more or equally advanced than other civilizations out there. And these civilizations might hope to find a sign of a cosmic neighbor just as we do.

If the other civilization is more advanced than we are, and given the age of the universe that is entirely possible, than we could learn a lot from them. This could provide all the sciences an enormous boost and therefore make us understand the universe better. It could also enrich the arts with works of art that no human has ever thought of. And it could lead to social or political systems that work better than what we currently have. There are also a lot of technological capabilities that we could gain from such a contact, from controlling the planetary climate to interstellar travel.

Or is making our presence known so risky that it would be a foolish idea to do it? In this case, as Alexander Zaitsev put it, *"Does the acronym 'SETI' deserve to be decoded as the Search for Extra-Terrestrial Idiots?"*[13], because it would be idiotic for anyone to transmit?

The opponents of METI are afraid of possible negative consequences to humanity. This ranges from fear of invasion by technologically advanced ETIs to negative cultural impacts that such contact may have.

Regarding an alien invasion, I wonder of what use we or our planet could be to an extraterrestrial civilization capable of interstellar travel. They would be obviously more advanced than we are if they can get here. So, it's not our technology they're after. The resources found on Earth can be found in large quantities anywhere else in the galaxy. Enslaving us is a stupid idea as we wouldn't be as obedient and capable for any form of manual labor as robots. And if they can travel over interstellar distances, they surely can build robots. Even we can do that. And they are getting better at doing much of our work almost by the day. And what if they want to study our biology and culture? Well, in that case, it would probably be better not to kill us all.

And if they are so much more advanced than we are, wouldn't they have found us already? After all, we have been emitting radiation for decades in the form of TV signals and radar, for instance. What's more, they could analyze our atmosphere from afar and see biosignatures. Even we are on the verge of being able to do this.

Still, an invasion is a possibility that cannot be entirely ruled out.

Opponents also fear the possible negative influences on culture and religion. This is, of course, actually an argument against SETI, not just METI. There might be some difficulties adapting to the new reality that we are not alone in the universe. But if the existence of extraterrestrial life is a fact, I'd rather know it and adapt my worldview accordingly than to live in ignorance of reality.

Being fearful of sending messages ourselves leads us to something called the SETI paradox. This term goes back to a paper by Alexander Zaitsev published in 2006. In essence, it states that we are expecting to receive signals from ETI but have a strong aversion to transmitting such signals ourselves. And if this is typical for all civilizations, SETI can never succeed. In other words, we are expecting others to do what we are not willing to do ourselves.

So, is it just fear of the unknown? Fear of change? Or does the risk really outweigh the benefit? And how do you decide that if there is no way you can realistically assess neither risk nor benefit because of too many unknowns?

Although famous astronomer Carl Sagan was not an active proponent of METI, in his popular book *Contact* and the movie of the same name, we are contacted only after they received a signal from us (in this case a TV broadcast).

So maybe we should do our part to end the great silence, instead of just sitting here and waiting for someone else to make the first move.

As I said, METI is a controversial subject. Both sides have their points. And considering the potential consequences of transmitting (or not transmitting!), it is not a decision that a single person, organization or state should make on their own. Luckily, there are serious efforts, most notably by *METI International* and *Breakthrough Message*, to start a global discussion on that topic. That's a good thing. We just need to keep in mind that we have to decide at some point and then act on it, whether that means staying silent or picking up the phone and start dialing.

Past attempts

There have been some attempts in the past to send messages to ETI (an overview can be found on Wikipedia[14]. None of them were approved by their national government or an intergovernmental organization such as the *United Nations*. But since there is no law against transmitting messages into space, no one stopped them.

All of these messages were controversial. But there haven't been any internationally coordinated, strong and long-term transmissions. This is, however, what any serious METI effort should look like. The *Arecibo-Message* of 1974, for example, was composed by just a few people and the transmission lasted for less than three minutes. If ETI doesn't happen to look our way for the three minutes when that message arrives, they will never hear it.

Some people would classify things like the *Voyager Golden Records*[15] as a part of METI. And this too is a message for ETI. But the speed at which these probes move is so slow compared to the speed of light at which our radio transmissions travel that they are not further considered here.

San Marino Scale

In 2005 Iván Almár proposed a scale to quantify the potential impact of transmissions from Earth. He presented his paper at a conference in San Marino, and hence it goes by the name *San Marino Scale*. The *San Marino Scale* is an ordinal scale between one (insignificant) and ten (extraordinary). This scale was officially adopted by the *SETI Permanent Committee* of the *International Academy of Astronautics* (*IAA*) in 2007.

On this scale, based on an evaluation by H. Paul Shuch and Iván Almár[16], the *Arecibo-Message* of 1974 would be considered an 8 (Far-Reaching). However, even our planetary radar used for analyzing potentially dangerous Asteroids would be deemed to be a 6 (Noteworthy). And that is undoubtedly something we do not want to stop using. After all, if the human civilization is annihilated by an asteroid impact, it doesn't matter if an ETI knows we were here.

Why haven't we seen them?

There is no way of writing a book about the Search for Extraterrestrial Intelligence without mentioning the *Fermi paradox*.

It is named after physicist Enrico Fermi, who in 1950, during a lunchtime conversation, formulated what is now known as the *Fermi paradox* to every member of the SETI community.

He reasoned that there is a massive amount of stars in the galaxy. If only a small fraction of them had a planet that is inhabited, then there should be lots of ETIs out there. Furthermore, he reasoned that our galaxy is way older than Earth. So ETIs would have had a lot more time to develop than humanity. Finally, if some of them developed interstellar travel, they should have been able to traverse the whole galaxy over the course of a few million years. But he saw no evidence of extraterrestrials. Which let him to the famous question "Where is everybody?"

There have been countless solutions proposed by various people. In fact, if you ask ten scientists on their take on the *Fermi paradox*, you will get ten opinions, all differing from each other, and partly or entirely contradicting each other.

I've listed a few of them below. They can broadly be divided into three categories: Hypotheses that state that we are rare, alone, and those that state that the galaxy is full of life, but we haven't seen any signs of it for various reasons.

Category I – we are rare

Hypotheses in this category presume the existence of what is called a "great filter." They suppose that there is some barrier to the development of any life form on its way to an advanced technological civilization that is capable of colonizing the galaxy. And this barrier is supposedly extremely unlikely to be overcome.

Possible reasons for our rarity are manifold. They range from a galactic predator-civilization (biological or artificial) that wipes out all civilizations that reach a certain technological threshold to the periodic extinction of life by natural events (like Gamma Ray Bursts).

Or maybe all or nearly all civilizations destroy themselves before reaching advanced technological capabilities like interstellar travel. They might wipe out their civilization through global war or nuclear weapons. Or they might develop an artificial intelligence that turns on them. Although in that case, the question remains: wouldn't we see signs of this artificial intelligence as there is no reason why it should seize existing after its creators are eliminated.

It's interesting how the event that is proposed to be a great filter depends on the time the hypothesis is published. Some of the proposed great filters are nuclear war, climate change or a super-virus that kills us all (or turns us into zombies). It seems to me like these explanations are more of a mirror of the dominant topics for humanity at the respective time of publication. Of course, these hypotheses are not impossible. But it shows how biased we are while speculating about the fate of other life in the universe.

We are not sure how life got its start on Earth. So it might very well be that the emergence of life is a process that is way more complex or unlikely than we currently think. And that therefore there is almost no life in the universe.

The emergence of life might be an improbable event. But the Earth is 4.54 billion years old. 130 million years later oceans formed. And another 130 million years after that, life first appeared on Earth. So, life arose on Earth 4.28 billion years ago, just 0.26 billion years after Earth formed. That happened remarkably fast for an unlikely event. But it doesn't rule out that it is unlikely. Especially as we only have one example (the Earth).

Maybe the evolution of multicellular life is a great filter. Then again this happened multiple times independently of each other on Earth (at least 46 times in eukaryotes alone[17]). This doesn't sound like an effective great filter.

There is also the possibility that life is common, but intelligence is not. The great filter might be somewhere between the emergence of life and the development of intelligence. Or at least intelligence that is capable of building technology, which is basically the difference between apes and humans. As far as we know, human-level intelligence capable of developing technology has only arisen once on Earth. Does that make it a good great filter? Maybe. We have no idea if intelligence is an inevitable outcome of evolution sooner or later. This is a controversial subject among scientists.

If life, for whatever reason, is rare in the universe than it's no wonder we haven't seen any signs of it so far. Space is huge. The distances between the stars are enormous. Should there really be no way of faster-than-light travel than it's not surprising we haven't seen or heard from anyone yet. By the way, if there is a way for faster-than-light travel, we probably wouldn't have noticed anyone either, because we have no clue how this would work. Also, let's not forget that the farther we look out from our Earth towards the stars, the farther we look back in time because light moves at a finite speed. If we look at a star that is 10,000 light-years away, we see it as it looked 10,000 years ago. Who is to say that there is no intelligent civilization on that planet right now?

Category II – we are alone

There is another possibility. We're not rare, we're alone. We're the only life, or at least intelligent technological life, that has developed so far.

Maybe the emergence of life is so incredibly unlikely that we are the only ones. Seems almost unbelievable given the size and age of the universe, but we cannot completely rule out this possibility.

A more encouraging alteration of this hypothesis is that we are the first or among the first because conditions in the universe have just recently become favorable to the development of life. That would mean that there are no advanced super-civilizations in the galaxy. But we, and also others, are on our way to that state. In this case, it would be only a matter of time before we finally meet each other.

Category III – we are one among many

There might be advanced ETIs. But none of them set out to colonize the whole galaxy, because there is no reason for them to do that. Maybe they've settled in a few star systems near their home planet, and that's it. Space is vast, and there are plenty of resources. There might be no good enough reason to colonize the whole galaxy. They might want to explore the rest of the galaxy or even the universe, but for this a few probes are sufficient. You don't need to colonize a star system to explore it. And such probes, as we have discussed in the section on SETA, probably wouldn't be noticeable by us, even if they were in our solar system. They also might not spread further out than a few star systems from their home because they have neighbors who do the same.

It is possible that there are many ETIs, but they're all listening, and no one is transmitting something for the others to receive. This is sometimes called the *SETI paradox*. It gets more and more unlikely the more ETIs there are. But from our point of view, it is a plausible explanation. After all, we're not deliberately transmitting much. There have been very few attempts at making our presence known and they've all been controversial. If the others are also not broadcasting because they fear it might be dangerous or a waste of resources, then we can listen as much as we like, we will never hear anything from the others.

Closely related, it could be that no one is transmitting, not because they fear it could be dangerous, but they know that it is. There might be some predator civilization(s) out there that annihilate any civilization they find. If this is the case, the ones that start transmitting are eliminated soon after doing so. And the ones that survive will keep quiet. Of course, it is possible that there is no predator civilization, but just the fear of one keeps everyone from transmitting. After all, this is what's happening on Earth. Measures to make our presence deliberately known are controversial precisely for this reason. The opponents of such efforts fear that there might be hostile ETIs out there, but they don't know.

Another thought seems quite scary. They might not want to explore the galaxy anymore once their entertainment technology is advanced enough. I find it scary because I see a tendency in humanity to go that route. They might have very advanced entertainment technology, like really advanced Virtual Reality. They might create a sort of *Matrix* and spend all their time in there, experiencing all they want with no interest in the real world around them anymore. They might have created incredibly advanced VR headsets or uploaded their consciousness directly into a computer. The result would be a complete halt in technological development and exploration of the universe around them. In this case, they would have no interest in talking to any other civilizations in their galactic neighborhood.

The *planetarium hypothesis* (proposed by Stephen Baxter) is similar, but this time it's the ETIs that have created the simulation for us. In this case, we wouldn't see any evidence of extraterrestrial life because an advanced ETI doesn't want us to see any evidence.

Then there's the *Zoo Hypothesis*. It states that there are extraterrestrial civilizations out there, but they are deliberately not contacting us so we can develop undisturbed. There might be a galactic rule that prohibits approaching a civilization below a certain threshold. That threshold may be technological, ethical, political or something else. If you watched *Star Trek*, you will recognize the *Federation's Prime Directive* as such a rule. The *Zoo Hypothesis* has been criticized as unlikely because it needs just one rogue civilization to violate such a rule and we would know about the existence of ETIs. But there are several ways around this problem. There could be an ancient super-civilization (maybe the first in our galaxy) that is way more advanced than the others and is able to enforce the rule effectively. Or there might be a galactic consensus between all civilizations, so nobody wants to break the rule because they all see its value. As a third option, there could be several larger factions in the galaxy. We might be in the territory of a federation (to use a notion from *Star Trek* again) that uses this rule. Even if another faction doesn't agree with them on that rule, why would they risk trouble to contact an underdeveloped civilization that not only lies within the borders of another faction but is also not of much use to them because of its low level of technological or scientific development?

Another possibility is that they are out there, but they haven't contacted us yet, because they don't yet know we exist. After all, there have been very few attempts to make our presence known to others in the past. And yes, there is leakage radiation from TV, radar, etc. that is leaving our planet. But these signals are very weak. And both deliberate and leaked signals move with the speed of light, which means they haven't gotten very far.

Tying into that, vast regions of the galaxy could be densely populated. But we happen to live in a neighborhood that is not or only very thinly populated and therefore of no great interest to most ETIs.

And then there is the possibility that they are out there and have already tried to contact us, but we weren't listening. Communication with radio waves hasn't been around until the end of the 19th century. Let alone lasers or the capability to detect neutrinos. They could have tried to contact us for 10,000 years straight. Without a detector for such a signal, we wouldn't have noticed anything. Also, keep in mind that data analysis is not done by hand but by computers. These computers were programmed to look for specific signals. But they might have missed something. If the computer sorted out an ETI signal because it didn't match the pre-programmed criteria, no human has seen the signal. In the future, artificial intelligence will likely bring a huge boost to what signals computers can detect.

Furthermore, as pointed out in the section *where should we look for signals?*, there are lots of places where we could look for ETI signals. And we're actually looking at just a small fraction of them and not even at the whole sky and also not all the time. The chances are significant that we might miss a signal just because we weren't looking at the right time, at the right spot in the sky, at the right frequency.

The same goes for technological artifacts of ETI, as pointed out in the section on SETA. There could be probes from an extraterrestrial civilization in our solar system right now without us knowing. Our current instruments are not nearly powerful enough to find small, dark probes in our own backyard. Also, we're not looking that hard. Our solar system is pretty big compared to a probe that might be just 1m in size, let alone nanoprobes, which would be a lot smaller.

They might use technology to communicate or travel that we haven't invented yet. And therefore we either see no signs of it or do see signs but are unable to interpret them as what they really are. For example, every Fast Radio Burst could be produced by a starship that activates faster-than-light propulsion. That is, of course, highly speculative and unlikely, but it illustrates the point.

Taking this line of reasoning a bit further, we might be too primitive to even fathom their undertakings. Thinking of the astounding speed of technological advancements in the recent decades, we will probably not even recognize our own descendants a thousand years from now. Now think about what a civilization might look like that is not thousands but millions of years ahead of us. They could be all around us, but we are too primitive to understand what they are and what they are doing. And they make no deliberate effort to inform us of their presence or even establish contact with us because we're to them what insects are to us. They just don't care to share their knowledge with us, and we're not capable of comprehending anything they do.

The list of possible explanations goes on and on. But I will leave it at that because, as with many things in SETI, we just don't know. We don't know how life got started in the first place. We don't know anything about the motivations of possible ETIs or the technology they would use. We cannot really imagine what a Kardashev Type III civilization would be capable of. We suspect that we would see evidence of large-scale constructions in, and alteration of, the galaxy if there were a civilization that would span the whole Milky Way. But that is not necessarily the case. Humans a hundred, let alone a thousand, years ago would not have dreamed about the technology that is completely normal to us now. If you were alive in 1990, did you expect your phone to be a small, powerful computer that is capable of taking photos and playing music, can be used as a navigation system anywhere in the world and lets you browse the internet wherever you are (not to mention what became of the internet) by the 2010s? And that was not long ago.

We can, and we should use our reason to speculate about solutions to the *Fermi paradox*. But we shouldn't be so arrogant to think that we can understand the intentions and capabilities of every civilization out there. They might very well use technology that is entirely unobservable or indistinguishable from natural phenomena to us with our current level of technology and understanding of physics.

We might not even recognize ourselves anymore a few hundred years in the future. There is a chance that humanity will reach *technological singularity*[18] in the next couple hundred years. Some reputable people even think it will happen within this century. How can we be sure to understand civilizations that are millions of years ahead of us in their development?

Most importantly, we can debate all day long about galactic colonization and why we don't see any evidence of extraterrestrial civilizations. But that must not keep us from actually looking for said evidence. The fact that the Earth is a sphere and not flat was discovered by observation and not mere speculation. In the 20th century, physicians thought that the human body could not function in weightlessness and astronauts would just die. That was until we flew people up there and saw that they were fine. Theory and experiment (or observation) go hand in hand to advance our understanding of the universe around us. We must not forget one over the other.

In the end, the discussion about the *Fermi paradox* will not be settled by debating. It will be solved by evidence found through scientific search efforts for signs of extraterrestrial life.

Even if intelligent technological life is rare or nonexistent, if there is more primitive life, we will probably find it over the course of the next decades. As for alien technology, that could be tomorrow or a thousand years from now. It all depends on our search efforts and the technology they use.

Kardashev Scale

The *Kardashev scale* is something that comes up again and again in discussions about extraterrestrial civilizations. So let me give you a quick overview.

In 1964 Russian astrophysicist Nikolai Kardashev proposed a scale to categorize extraterrestrial civilizations based on their energy usage.

He divided civilizations into three categories:

A Type I civilization, also called a planetary civilization, can use all the energy of its planet.

A Type II civilization, also called a stellar civilization, can use all the energy from its star.

A Type III civilization, also called a galactic civilization, can use all the energy of its galaxy.

Numerous alterations to the original scale have been proposed. These range from expanding it up to Type V, which would be capable of using the energy of multiple universes, to basing it on different metrics, as the original *Kardashev scale* is based solely on energy usage of the civilization.

What happens if we find something?

We've been pointing our telescopes up at the sky, looking for that one signal. The one that finally tells us "you are not alone." One day, a SETI researcher sees an interesting signal on his screen. But what now?

Confirmation of a signal

Picking up a signal of an extraterrestrial civilization would be amazing. But we have to be sure that it's actually an ETI signal.

If an individual or institution picks up a signal that seems to be of artificial extraterrestrial origin (a so-called candidate signal), it should first try to confirm that it is indeed of artificial and extraterrestrial origin.

Extraterrestrial

It has to be confirmed that the signal originates from a place that is not on Earth. Quite regularly, signals from terrestrial transmitters, called RFI (Radio Frequency Interference), are picked up by telescopes. We might, for instance, pick up a radar signal.

If we use a telescope that tracks a point in the sky (like a specific star), then we move the telescope to another location in the sky. If the signal is still there, it is not coming from the point in the sky we were just observing.

If we use a telescope that does not track a specific target, we just wait. The Earth rotates and will take care of pointing your telescope somewhere else. In the case of the *Arecibo Radio Telescope*, for instance, we would typically see a signal getting stronger and then weaker again over a 12 second period. That's because over that period a signal from a specific point in the sky comes into the telescopes field of view at the edge of the dish, gets stronger as it moves towards the center, and then gets weaker again as it progresses from the center out of the field of view at the opposite edge. All due to the rotation of the Earth. A signal that does not show this pattern is probably from Earth.

Another test (and a pretty reliable one) is to use a second telescope. If we point a second telescope at the same point in the sky and also find the signal with the second telescope, that's a good indication that the signal is of extraterrestrial origin. The farther away the telescopes are from each other, the better. After all, a second telescope at the same site could be picking up the same RFI. If we use a telescope that is, say, on another continent, this is an excellent indicator that the signal is indeed not from Earth.

Artificial

We have to find out if the signal is not just a natural phenomenon. We know a lot of natural events that produce radio waves. But even if it is a signal that we don't recognize as a known natural phenomenon doesn't necessarily mean it's artificial. It is pretty much certain that there are things out there we don't know about yet. There might be some astrophysical process we don't know yet that produces a signal which, at first glance, might look artificial.

A good example is the discovery of pulsars. The first one was detected in 1967, and the signal was actually a regularly pulsed radio signal. Such a signal might be considered artificial. And in fact, the discoverers Jocelyn Bell and Antony Hewish at first thought it might be a signal from an extraterrestrial civilization (hence the informal name LGM 1 for "Little Green Men 1"). Today, we know that these signals come from pulsars, which is a particular form of rotating neutron star. So, if we were to discover a signal that might, at first glance, look artificial doesn't necessarily mean it is. Just because we don't know something is not viable proof that it's artificial.

As we cannot really know what a signal from an extraterrestrial civilization would look like, we have to make some assumptions. There is no universal consent in the SETI community on which signals to look for. One hypothesis that is widely used for SETI efforts is the assumption that ETI signals would have a narrow bandwidth. These kinds of signals would be easier to filter out from the noise of natural sources because no known phenomenon produces signals with a really narrow bandwidth. All known natural sources radiate over many frequencies.

Furthermore, it would be more economical to send a signal with a narrow bandwidth. The more frequencies you are transmitting at simultaneously, the more energy you need. Therefore, a lot of searches focus on narrowband signals. It should be noted, however, that there are some efforts to look for broadband signals. *Astropulse* is such a project.

Another good indicator that the signal is artificial is if it contains encoded information. If we would be capable of decoding their message and interpret the information they're sending (e.g., a sequence of prime numbers or the number pi) that would undoubtedly be proof of artificiality. But even if we were unable to interpret the information received we might be able to recognize the encoding. If you want to send a digital message you have to use some form of encoding. They will hopefully be sending messages that are easy for us to decipher and understand. Humans have theorized that simple mathematical concepts would be an excellent base to start and establish a higher form of communication from there. Maybe ETIs have similar reasoning. But even if we cannot understand the message, the encoding that was used might have patterns to it that look utterly non-natural. So, even if we don't understand the message we might be able to see that there is a message. That, of course, would also be excellent proof of an artificial origin.

Post-detection cover-up

The assumption that any message from an extraterrestrial civilization would immediately be covered up by some government persists even at the time of this writing, which is the year 2018.

This scenario, however, is unlikely due to a number of factors.

SETI research is not carried out by government agencies. Although *NASA* once had a program to search for signals of artificial extraterrestrial origin, those days are long gone. There is no serious effort to search for ETI signals at *NASA* or any other governmental space agency today. This, in turn, means that SETI research is carried out by private institutions. And these are not as easily controlled by a government as a government agency is.

The instruments used for SETI are located in different countries throughout the world. And the government of one country cannot decide what to do with the data gathered at a telescope in another country.

People are on social media. Although it is commonly accepted that a signal should be confirmed by other sources before the release of the information to the public, it seems likely that someone on the team of researchers who discovered a signal might post something on some social network. It doesn't even have to be one of the researchers. Anyone in the facility where the message has been detected could leak some information. Once information of an ETI signal is on a social network, this news will spread like wildfire.

Post-detection principles. Major players in the field of SETI have agreed to adhere to some principles following the detection of a signal of artificial extraterrestrial origin. Although these principles have no power of law, many reputable institutions and individuals have agreed to a declaration called *Concerning Activities Following the Detection of Extraterrestrial Intelligence*. In this declaration is not a word to be found about hiding the fact that a message has been received.

Continuous observation. If we pick up a signal from an extraterrestrial civilization, we surely don't want to lose any of the information that is sent towards Earth. But the Earth rotates. This means, to continuously observe the signal we have to use telescopes around the world. We would have to inform other institutions so that they look at the right part of the sky with the right instrument on the right frequency.

Anyone else could pick up the signal. If a message is sent towards the Earth, it will not be detectable by one country alone. Telescopes are built around the world in many different countries. If a national government tries to cover up the detection of an ETI signal with one of its telescopes, it is very well possible that the same signal will be detected by another telescope in another country simultaneously or shortly thereafter.

Cultural impact of extraterrestrial contact

The probability of the detection of an ETI signal with our current technology cannot be estimated because of too many uncertain factors. But if we were to find a signal that is undoubtedly of artificial origin, it seems very likely this would have a profound impact on society. And not just on one country or one culture or one religion, but on all humans on Earth (and a few that might be off the Earth at that time).

The societal impact, of course, heavily depends on the intentions and capabilities of the ETIs. Are they friendly or hostile? Do they offer knowledge? Do they offer guidance or coercion? Are they capable of getting here?

There are a whole lot of hypotheses on the societal effects that contact with ETIs might have. They range from destruction of human society to the rise of Utopia and anything in between.

Over the last years, it seems to be a universal trend (at least in western cultures) to paint our present-day world dark and the future even darker, while simultaneously romanticizing the past. Pretty much every TV show or movie of the last years that plays in the future portraits a dystopian future. Apparently, stories like that sell well, but what do they do with our vision of the future and our expectations for the future?

What happens to your worldview if all you see is negative news of the present and dystopian visions of the future? This is, in my view, a big cultural problem of our day. Do you think people will expect ETIs to be friendly or hostile when they are bombarded with negativity every day?

Although this perception of the world around us is a big problem, it is still just our perception. If we look at the real numbers, things seem very different. There has been some effort to objectively analyze how our world has developed over time. Especially noteworthy are the works of Steven Pinker and Hans Rosling. And these studies consistently show that the world is getting better over time. It is better than it was in the past in pretty much any regard imaginable. And it seems this trend will continue.

Our own history is often referenced to demonstrate that an encounter with ETIs would be disastrous for us as the less advanced civilization. However, these comparisons usually refer to single events in our history. If you don't just look at individual events, but rather our development as a species and civilization, the picture doesn't seem so grim anymore. There are less authoritarian regimes, fewer wars, less violence in general and our perspective becomes more and more global. Seeing how we as humans and society as a whole developed over time, I'm hopeful that more advanced ETIs will not be hostile to us. And less than or equally advanced civilizations won't be able to do us any harm, as they wouldn't have the technological capability to do so.

It is also entirely plausible that, due to recent advances in the detection of exoplanets and the resulting media coverage, many humans would not be surprised if we were to detect an ETI signal. And that, as a consequence, there would not be a lot of effects at all on human society.

The societal impact of contact with an ETI, be it via a signal or physical presence, depends on different factors. One attempt to measure the potential effects of an ETI signal is the *Rio Scale*. It gives a score from 0 - 10 for the impact a detection would have on society. It takes into account the class of the phenomenon (ET artifact, beacon, physical encounter, etc.), the type of discovery (from archival data, steady phenomenon, etc.), the apparent distance to the source (within the solar system, extragalactic, etc.) and the credibility of the report.

The *Rio Scale*, although officially adopted by the *IAA* (*International Academy of Astronautics*) *SETI Permanent Committee*, has been criticized as being oversimplifying.

It seems likely to me that if a signal is discovered, the discoverer will want to share this news, regardless of the value on the *Rio* or any other scale. So I think the *Rio Scale* is just what the *IAA* calls it: A work in progress *"in order to bring some objectivity to the otherwise subjective interpretation of any claimed ETI detection."*[19] But in the case of an actual ETI detection, it will probably be just briefly considered or outright ignored before the news is made public.

In the end, the possible effects on human society are manifold and almost impossible to predict. There has been a lot of theorizing about the impact that contact or even knowledge of the existence of an extraterrestrial civilization could have on society on different levels (theological, political, etc.). But however reasonable these hypotheses might sound, we would be wise to keep in mind that all of the predictions and assumptions are made by humans and therefore are heavily biased. Of course, we cannot help but look at this question through our own eyes. After all, we don't know any other perspective than the human one. But we should keep in mind that every possible scenario we come up with is biased by our history and our current worldview. That applies to the societal effects of an ETI contact as well as to possible motives and behaviors of the ETIs.

Projects

We will now take a look at past and present projects to answer the question "Are we alone?"

What have we done so far?

There is no way to list any and all SETI projects in history. Numbers vary from author to author, and not all projects were brought to the attention of the public. There were projects in different countries and also lots of private projects. So, I listed just a few popular ones in the following section of the book to give a sense of what has been done in the past.

19th and beginning 20th century

In the 19th century and the beginning 20th century, it was widely believed that there is life on other bodies of our solar system, especially on Mars, the Moon and Venus. During that time, there were quite some scientists seriously thinking about how to communicate with these other civilizations. There was no lack of proposals.

A few examples:

- In 1818 the great mathematician Carl Friedrich Gauß suggested using his Heliotrope. This was an instrument invented by him to measure areas accurately. It reflects sunlight to a point that can be precisely targeted. His suggestion was to build a giant Heliotrope and target the Moon with it so that life forms on the Moon would notice the signal and therefore know that there is another civilization on Earth.
- A few years later the Austrian astronomer Joseph Johann von Littrow suggested digging long channels building geometric forms in the Sahara desert, filling them with water and Kerosene and then lighting them up during nighttime.
- In 1875 Finnish mathematician Edvard Engelbert Neovius published his idea to construct a system of an enormous amount of lamps in the South-American Andes to communicate with Martians.

None of these or any other proposals were built, mainly due to financial limitations.

In 1867 James Clerk Maxwell theoretically predicted the existence of radio waves. In the 1880s Heinrich Hertz produced radio waves in his lab and thereby proved Maxwell's theory (it was actually more generally about electromagnetic waves, not just radio waves).

As it became clear that it is possible to transfer information with radio waves in the following years, it didn't take long until people had the idea to use it for interplanetary communication. To the proponents of this idea belong popular names like the famous radio pioneer Guglielmo Marconi, astronomer David Peck Todd and Lord Kelvin (aka William Thomson) who all stated their belief that Martians could be contacted via radio waves.

Nikola Tesla not just believed that a Martian civilization could be contacted with his "Wireless Electrical Transmission System." In later years he became convinced that he had picked up a transmission from Mars. Although it is not completely clear what Tesla actually picked up, it seems rather obvious to us now that it was not a signal from a Martian civilization.

There was even a *National Radio Silence Day* in the U.S. in 1924. It was established to be able to listen to possible incoming signals from Mars without interference from our own radio transmissions. The most powerful transmitters of the *U.S. Government* were temporarily turned off during a three day period. However, most private broadcasting stations didn't care, and it seems likely that the military also kept using their transmitters. As you might guess, no signals from a Martian civilization were detected in 1924.

The beginning of modern SETI

There were a lot of ideas on how to communicate with extraterrestrials in the 19th and beginning 20th century. And there were even some attempts to listen to signals from civilizations outside of Earth. But the first systematic, scientific search for signals of intelligent extraterrestrial origin did not take place until the second half of the 20th century.

SETI as we know it today started with an article titled *Searching for Interstellar Communications* by Philip

Morrison and Giuseppe Cocconi published in the magazine *Nature* in 1959. In this now famous article (at least in the SETI world) they not only proposed to systematically search for signs of interstellar communication but also suggested doing so in the microwave part of the spectrum, specifically around the frequency of 1.42 GHz. This paper and especially the frequency of 1.42 GHz had a massive influence on the modern field of SETI and are still relevant to this day.

During that time, the whole idea of SETI was not very popular. But the fact that two renowned physicists published a scientific article on the topic in a respected magazine like *Nature* gave the whole idea more credibility.

After Morrison and Cocconi laid the theoretical groundwork, American astronomer Frank Drake was the first to actually do a search with the so-called *Project Ozma*. It seems worth noticing that Drake had already begun to plan *Project Ozma* before the publication of the article by Morrison and Cocconi. During the preparation, he even came up with the idea to search around the frequency of 1.42 GHz. However, Drake did not publish any details about his project before the publication of the article in *Nature*. And so, Morrison and Cocconi are usually credited with coming up with the idea.

In 1960, Drake started *Project Ozma*, the first systematic scan of different target stars for signals from intelligent extraterrestrial life forms. To do that, he used a 26 meter (85 ft) telescope in Green Bank, West Virginia, USA. He looked at the stars Tau Ceti and Epsilon Eridani. With this telescope, he was able to observe just a single channel of 100 Hz bandwidth at any one time. He found no convincing ETI signals.

After *Project Ozma*, there were several search programs carried out. The exact number is uncertain as numbers vary from author to author. What is certain is that Soviet scientists also did a number of searches in the time after *Project Ozma*.

Big Ear SETI

Dr. Robert S. Dixon, then Assistant Director at the *Ohio State University Radio Observatory*, phrased it wonderfully, so I just let him speak[20]:

> *"During my project Cyclops research, it became clear to me that many theoretical papers were being written about SETI, but nobody was doing any extensive actual searching."*

Out of this realization came the first full-time SETI program in history, conducted at the Ohio State "*Big Ear*" radio telescope. It ran from 1973 until 1995 and is still the longest running SETI program in history.

It was during this program that the famous *Wow! Signal* was found. In 1977, the *Big Ear* radio telescope received an interesting signal. It was the kind of signal we would expect to see if an ETI is trying to call us. It was strong and narrowband (meaning it was confined to a specific frequency or small frequency range). Narrowband signals are not something we usually see from natural sources.

Furthermore, it was near the "magic" frequency of 1,420 MHz. Its source must have been at least at lunar distance (the distance of the Moon). Which means it did not originate from the ground or satellites in Earth orbit.

It was detected and recorded. But this was the time when received signals were printed out on paper and had to be analyzed by hand. So it was not discovered until a few days later when astronomer Jerry Ehman reviewed the piles of paper generated over the previous days.

He found the signal and was so impressed by its strength that he wrote "Wow!" next to it on the paper. And this is where the name of the signal came from. Of course, Ehman and others tried to find that signal again. But unfortunately, they weren't able to find it anymore. And despite many attempts by various researchers over the years, it was never seen again.

What was it? Was it a secret military satellite? Or an unknown natural phenomenon? Or was it a signal from ETI? Nobody knows for sure. Although definitely interesting and possibly an ETI signal, it cannot be regarded as evidence of extraterrestrial intelligence because it was never observed again. And there is no way to gain any more information from the original signal than we already have. We cannot look for any encoded information in the signal because the records we have only show the strength of the signal and its frequency and nothing more.

1980s and beyond

During the 1980s and in the years that followed, several search programs were conducted by different actors. A few of them are listed below.

In the 1980s the Harvard University carried out the *SENTINEL* program.

During that time in Europe, the *Nançay* radio telescope in France and the 32 m radio telescope in Medicina, Italy were used for SETI projects.

In 1985 the *Megachannel Extra-Terrestrial Assay* (*META*) was launched. This led to the *Billion-channel Extra-Terrestrial Assay* (*BETA*) in 1995. Both *META* and *BETA* used the *Oak Ridge Observatory* in Harvard. Unfortunately, it was severely damaged by a storm in 1999. This meant the end of the *BETA* project.

In 1992 *NASA* launched a SETI project called *Microwave Observing Program* (*MOP*). It was intentionally given such an unobtrusive name because *NASA* feared intervention from politicians if they just called it what it actually was, a search for signals from extraterrestrial civilizations. As it turns out, their fear was not unfounded. The *US Congress* still found out about the program, and it was canceled just one year after its start.

Unwilling to give up on the search for extraterrestrial intelligence, the privately funded *SETI Institute* took up where *MOP* left off. Under the direction of Jill Tarter, they started *Project Phoenix*. It ran, with interruptions due to lack of money and booked out telescopes, from 1995 - 2004. They actually put all their equipment in a container and traveled from one telescope to the next. All in all, they observed about 800 stars up to a distance of 200 light-years.

From 1997 and 2007 the *Berkeley SETI Research Center*, the ones who are running *SERENDIP,* also ran a program called *SEVENDIP* (*Search for Extraterrestrial Visible Emissions from Nearby Developed Intelligent Populations*). Contrary to the other search programs mentioned here, this was one of the few searches in the optical part of the spectrum.

None of these projects, and none of the ones not mentioned here found a convincing signal of extraterrestrial intelligent origin.

SETI Live

In 2009, astronomer Jill Tarter won the *TED Prize*, which is endowed with $100,000.

Out of that came the *SETI Live* project. It let anyone with an internet connection analyze data from the *Allen Telescope Array* (*ATA*) in Hat Creek, California in near real-time.

Most data from the *ATA* is analyzed automatically. But there are frequencies that the computers are not good at analyzing. These are frequencies that have a lot of terrestrial interference (e. g., from ground-based radar to satellites in orbit). It was hoped that humans would be far better in finding possible ETI signals in these frequency bands. The project was canceled in 2014 due to lack of financial resources.

Fortunately, artificial intelligence will very likely significantly improve our computers' ability to find signals in the near future.

ATA

When you drive up to Hat Creek Observatory, about 300 miles northeast of San Francisco, you will see a field of 42 dishes, each 6.1m in diameter. The dishes look a bit odd, not like your typical satellite dish to receive TV signals. That's because they use a different design, called an offset Gregorian system. These dishes are radio telescopes, and together the 42 of them form the *Allen Telescope Array*.

The *ATA* was built mainly by the *SETI Institute* and is used for SETI research every day of the week. It isn't available 24 hours a day for SETI though. Other parties use the *ATA* for radio astronomy and satellite tracking to cover the substantial operating costs of the array. Still, this array is a considerable advantage for SETI. In the past, SETI researchers used existing radio telescopes. But observing time on these telescopes is necessarily highly restricted due to a large number of research projects that apply for observing time on these telescopes. The *ATA*, on the other hand, can be used every day for SETI observations.

Another advantage for SETI is that this array, contrary to single-dish telescopes, is able to look at multiple targets in the sky simultaneously.

As of now, it is unclear when, if ever, the array will be expanded to its planned size of 350 dishes, due to lack of funding. But the *ATA* is built in a way that it can be easily upgraded as technology improves and additional funding is provided.

What are we doing right now?

Several SETI projects are still ongoing. As with past projects I picked a few popular ones for this section of the book.

SETI@home

To find a signal of an extraterrestrial civilization, it is not enough to point a radio telescope somewhere in the sky and receive everything that comes in. There is a lot of data coming in. To find an interesting signal you have to analyze the data. In 1995, when it was common to use supercomputers to analyze the massive amounts of data, David Gedye, Woody Sullivan, Dan Werthimer and David Anderson began working on a project to use home computers, connected over the internet, as a "virtual supercomputer." This project was *SETI@home*, now called *SETI@home Classic*. It was based at the *Space Sciences Laboratory (SSL)* of the *University of California, Berkeley*.

The idea behind it was to take the data collected from radio telescopes (the *Arecibo Radio Telescope* at the time) and divide it into chunks that are manageable for a home computer (the so-called "work units"). Then the work units were sent to the participating home computers together with instructions on how to analyze the data. The computer did the calculations as instructed and reported all interesting signals that it might have found back to the servers in Berkeley.

In May of 1999, they released the software for the individual computers (the desktop clients) to the world. The demand exceeded their expectations, and hence they had some performance problems in the beginning. So many people downloaded the client software and ran calculations that their server infrastructure in Berkeley couldn't keep up.

After various improvements over the years, *SETI@home Classic* came to a close in December 2005. But that didn't mark the end of *SETI@home*. The *SSL* transitioned to *BOINC* (*Berkeley Open Infrastructure for Network Computing*). This is mostly an improved and more generalized version of the original *SETI@home* application. It is not specific to *SETI@home* anymore. There are numerous other projects now which use distributed computing just like *SETI@home* does, but for other purposes. There are projects to find pulsars, find cures for diseases and so on.

SETI@home wasn't the first distributed computing project (the *Great Internet Mersenne Prime Search* and *Distributed.net* came first), but it was (and is) a hugely popular project that gave rise to *BOINC*, which is now used by many other scientific (and even commercial) projects.

Arecibo

Have you seen the movie *Contact*? (If not, you definitely should) Remember that colossal radio telescope surrounded by beautiful nature? That wasn't a set, it's the *Arecibo Radio Telescope*. It was built into a natural sinkhole in Puerto Rico, USA.

Its dish has a diameter of 305m and is made up of more than 38,000 individual aluminum panels. 150m above this dish hangs the receiver, suspended from three pillars standing around the dish.

As the dish is not movable, the only way to point the telescope is to move the receiver. That way the *Arecibo* telescope can track celestial objects, but only for 2.6 hours at a time. After this time the rotation of the Earth has turned the telescope so far away that it cannot see the object anymore.

The *SETI Institute's Project Phoenix* came here in 1998. At the time it was the world's largest radio telescope, and the demand was correspondingly high. Therefore, *Project Phoenix* observed in multiple sessions, two of them per year, until 2004.

The *Arecibo* telescope was also the source of data for the *SETI@home* project for many years. The *Berkeley*

> *SETI Research Center* still gathers data there for *SETI@home*, but recently there has been a big influx of data to *SETI@home* from the *Green Bank Telescope*, collected as part of the *Breakthrough Listen* program.
>
> Also noteworthy is the so-called *Arecibo Message*. In 1974, Frank Drake and others used the telescope not to listen for ETI signals but to send a message of our own out into space. It should have been aimed at the globular cluster M13 about 25,000 light-years from Earth. Unfortunately, there was an error in the calculation, and the transmission will not reach its intended target. But I'm not too sad about that as it would have reached its destination no earlier than 25,000 years from now anyway. And I really hope that we will find proof of ETI way earlier than that.

SERENDIP

SERENDIP (*Search for Extraterrestrial Radio Emissions from Nearby Developed Intelligent Populations*) is a project run by the *Berkeley SETI Research Center* at the *University of California, Berkeley*.

It started in 1979 and is, with many improvements over the years, still running today. It currently uses the *Arecibo Radio Telescope* in Puerto Rico and the *Green Bank Telescope* in West Virginia, USA to gather data.

It uses a technique called "piggybacking" to gather a lot of data without having to buy dedicated telescope time. What this means is they just plug their devices in and record data while other researchers are using the telescope for their purposes. That does mean they have to look wherever the researcher using the telescope right now wants to look. But it allows them to get a lot of observing time almost for free.

Project Argus

The *SETI League's project ARGUS* aims at deploying 5,000 small radio telescopes around the globe to enable an all-sky survey 24/7. What makes this project special is that these are all radio telescopes built and owned mostly by amateur astronomers. These *Argus stations* are coordinated via the internet.

Project ARGUS started operations in 1996 with five telescopes and has been adding more of them since then. The sensitivity of these telescopes is, of course, a lot lower than what professional astronomers usually use. But nevertheless, this network of small telescopes would allow the first all-sky all-the-time survey just for SETI.

The last status report to be found on their website[21] is from 2005 when the project seemed to be well underway as planned and aiming for completion (5,000 stations) in 2020. It's been some time since then, but their website is maintained and (apart from the progress reports on *Argus*) regularly updated. So let's hope they'll meet their goal.

Breakthrough Initiatives

The *Breakthrough Initiatives*, founded in 2015 by Yuri Milner, are a set of programs all relevant to the search for life in the universe. As of this writing, there are five different programs.

Listen

Breakthrough Listen will spend $100 million on searching the 1,000,000 nearest stars to Earth for ETI signals at radio and optical frequencies. Furthermore, it will scan the plane and the center of our galaxy and the 100 nearest galaxies.

This is not just the most comprehensive SETI search in history. Funding has always been a problem for SETI efforts. With this amount of financial resources, the program can utilize some of the world's best telescopes, including the 100 m *Robert C. Byrd Green Bank Telescope* (radio), the 64 m *Parkes telescope* in Australia (radio) and the *Automated Planet Finder* telescope at *Lick Observatory* (optical). For the future, it is also planned to use the *SKA* (*Square Kilometre Array*), which, when finished, will be by far the largest radio telescope array ever build.

Message

Breakthrough Message is designed as a competition to develop a message that represents humanity and planet Earth, and that could be understood by an extraterrestrial civilization. It is open to everyone and comes with a $1 million prize pool for the best messages. But its purpose is not just to get a bunch of messages that could be sent to ETIs. Instead, it's *"to encourage humanity to think together as one world, and to spark public debate about the ethics of sending messages beyond Earth"*[22].

Watch

Breakthrough Watch aims at finding and analyzing exoplanets. It will spend several million dollars to search for Earth-like planets around stars within 20 light-years of Earth and look for biosignatures.

Starshot

Breakthrough Starshot is a $100 million program which aims to establish the technical capability to send small spacecraft at 20% of the speed of light to our nearest star system, Alpha Centauri, within a generation. Ideally, these probes would make a flyby of the recently discovered planet Proxima b and send pictures of it back to Earth.

This program is vital. With SETI we sometimes seem to forget that we have found life in the universe already. And that life is us. Although the discovery of thousands of exoplanets in recent years makes it seem likely that there is plenty of other life out there, it nevertheless is a possibility that we are the only (intelligent) life form, at least in our galaxy. I think that listening and looking for ETIs is important. If they are out there, I want to know. But if they are not, if we are the first, it is our job to spread life, our life, across the galaxy.

And even if there are others out there, it would still be a good idea to develop the capability for interstellar travel. There is so much more to explore and learn out there. We have to get out there eventually if we want to find it. Besides, actually traveling to other planets and stars, even if via a probe, seems far more exciting to me than just staying on Earth and watching at the sky.

Discuss

Breakthrough Discuss is an annual conference *"focused on life in the Universe and novel ideas for space exploration"*[23].

With all these wonderful programs to advance the search for life in the universe coupled with technological advancements over the recent decades and a general rise of interest in the fields of SETI and exoplanets, our chances of finding life elsewhere in the universe have never been better.

> ### GBT
>
> In West Virginia, USA there is a place where radio transmissions are strictly limited. On the premises of the facility we're now visiting, they don't even like vehicles with spark plugs or microwave ovens.
>
> If you happen to live within a 20-mile radius of this facility and plug in a Wifi-Router, you will get a visit from a patrol car that is actively monitoring the area for devices emitting significant amounts of electromagnetic radiation.
>
> Here, with green hills in the background, we find several radio telescopes towering above the green meadows. One of them clearly stands out because of its size. It is the *Robert C. Byrd Green Bank Telescope* (*GBT*), the largest fully steerable radio telescope in the world.
>
> You might notice some similarity in the design of the telescope with another telescope we already took a look at, namely the *Allen Telescope Array*. The receiver here is also not in the center but slightly offset to the side. The *GBT* uses the same offset Gregorian system. It is just a lot bigger with a diameter of 100m.
>
> The connection between Green Bank and SETI dates back to the very beginning of modern SETI. Back in 1960, Frank Drake conducted *Project Ozma*, the first systematic search program for ETI, here in Green Bank.

He didn't use the *GBT* for this project, however. The *GBT* wasn't even built at that time. He used a 26m telescope that is located just a short distance away from the *GBT*.

Project Phoenix also was at the *Green Bank Observatory* from 1996 to 1998. At that time they also couldn't take advantage of the massive *GBT* because it began regular science operations no earlier than 2001.

The *GBT* is, however, heavily involved with SETI today because it is one of the major telescopes used for *Breakthrough Listen*.

Jodrell Bank

When you come to the *Lovell Telescope* near Manchester in the United Kingdom, it looks just a bit like a SciFi space cannon. And indeed parts of the telescope come from two *World War I* battleships. And it is capable of sending out radiation. Although that is more like sending signals to our probes than shooting interplanetary death rays.

The *Lovell Telescope* is a fully steerable single-dish telescope with a size of 76m. That makes it still the third largest fully steerable radio telescope in the world as of 2018, more than 60 years after its construction.

The *Lovell Telescope* was used as a secondary telescope for *Project Phoenix* to confirm signals while they were observing at *Arecibo* (98-04). More recently it was announced that the observatory would conduct an independent SETI program in collaboration with *Breakthrough Listen*.

The *Lovell Telescope*, together with two other smaller telescopes, is part of the *Jodrell Bank Observatory*. This facility was chosen as the headquarters for the organization building the next telescope that we're going to take a look at, the *SKA*.

SKA

The *Square Kilometre Array* (*SKA*), if built, will be the biggest and most sensitive radio telescope ever constructed.

The *SKA* will be an array of a lot of different radio antennas. It will have two core regions. One in South Africa's Karoo desert for the high and mid frequency dishes. The other in Australia's Murchison Shire for the low-frequency antennas. The completed array will have antennas spread out over 3,000 kilometers (1,860 miles).

The *SKA* will be so large that its construction is divided into two phases. Phase 1, which should be completed in the mid-2020s, aims to build about 200 dishes in South Africa and roughly 130,000 low-frequency antennas in Australia.

Phase 2, which is planned to be finished in 2030, will increase that number 10 times. If built it would consist of an unbelievable number of 2,000 dishes across the African Continent and about a million antennas in Australia. That would give this instrument the highest resolution of any telescope, not just radio telescopes. The total collecting area of this array would be about one square kilometer, hence the name. If it actually will be built completely remains to be seen. But so far everything looks good.

If built, this telescope would be a gigantic step forward for SETI research. If there is a civilization with a level of technological development similar to our own in our stellar neighborhood, we will find them with this telescope, if they deliberately try to contact us or not. The *SKA* will be sensitive enough to pick up television broadcasts or an airport radar from a planet dozens of light-years away.

APF

On top of Mt. Hamilton, east of San Jose, California, at an elevation of almost 1,300m is our next stop. It is the *Lick Observatory*. Or more specifically the newest telescope at this facility, namely the *Automated Planet Finder (APF)*.

The *APF* is a robotic 2.4m optical telescope. Its primary purpose, as the name suggests, is to find planets around other stars. Specifically, it's designed to detect relatively small rocky planets similar to Earth by observing a list of nearby stars every night for months in a row.

That alone would make it interesting for the search for life in the universe. But with the help of *Breakthrough Listen* it is now also searching for laser signals from ETI. In fact, it is sensitive enough to detect a laser with no more power than a conventional household light bulb of 100W from a planet orbiting a nearby star.

Parkes

From Sydney, Australia, we turn our gaze to the northwest to a place of great importance to SETI. After a four and a half hour ride in our car, we arrive at the rural town of Parkes. After another 20 minute drive up to the north, you better turn your cell phone off, so you don't mess up the measurements. Here before us, we see a round building with a 64m dish on top of it. We've arrived at the *Parkes Observatory*, Australia's premier radio telescope. If we're lucky, we can see the four 15hp motors moving "The Dish," as it is sometimes called, towards its next astronomical object of interest. The telescope could do a full circle in 15 minutes at full speed. The movable part of the telescope weighs 1,000 tons, and that is enough to keep it in place. It's not fixed to the concrete structure below; its own weight holds it down.

The electricity supply here can be unstable at times. But not to worry, the facility has a set of 40 car batteries to keep the data save until the backup diesel generator takes over.

This telescope is famous for various scientific discoveries and its role in the *Apollo* missions, receiving signals when the moon was on the Australian side of the Earth. It made its "first contact" with SETI more than two decades ago when a rather famous researcher in the field of SETI visited the telescope. In February of 1995, Jill Tarter and her team of *Project Phoenix* came here to search for signs of extraterrestrial intelligence. They spend a few months here searching for artificial signals from 200 nearby stars that could only be seen from the Southern Hemisphere.

But that was not the end of SETI research at *Parkes*. The telescope is one of two major telescopes currently used by the *Breakthrough Listen* program (the other one being the *Green Bank Observatory*).

How can I get involved?

If you read this far, it's clear that you are interested in SETI. I hope you liked my book and got something out of it. But now the question remains: what now? What do you do from here? If you think SETI or METI or the search for extraterrestrial life in general is a worthwhile endeavor, how can you get involved?

If you want to get seriously involved in the search for extraterrestrial life, I suggest you start off by finding out what specific field interests you the most. Go ahead and get some books or take a few courses on radio or optical astronomy, astrobiology, exoplanets or other fields that might lead to the detection of an ETI signal like neutrinos, gravitational waves, etc.

If you're thinking about building a whole career on the search for extraterrestrial life: test before you invest! You better take a closer look at your options and then choose the research field that is the best fit for you. You could then get a degree in that field and start a traditional career at a university or other research institution or a national space agency.

You could also become an astronaut. Who knows? You might be the one discovering life on Mars! But be aware that the search for life will probably not be the primary focus of your work as an astronaut.

But you don't have to work at a national space agency to work on space-related stuff! There are plenty of private companies nowadays that have some relation to space. Although these companies might not directly be searching for ETIs, a lot of them are trying to make access to space easier and cheaper in some way. And easier access to space benefits the search for extraterrestrial life.

Two examples:

- *SpaceX* not only wants to lower the costs of going to space (as do others like *Blue Origin*). But this company has clearly stated the colonization of Mars as a long-term goal. Now, what better way to search for signs of past or present life on Mars then by sending a lot of people there? Our rovers do an amazing job, but a human could do all that way faster.
- *Planetary Resources* and *Deep Space Industries* want to mine asteroids, so we could refuel or even build ships in space without having to lift all that material up from the ground (which is very expensive). This will likely make the exploration of our solar system (and hopefully eventually other star systems) easier and cheaper to do.

But these are just examples of quite popular companies. There are a lot of other private companies to be found where you might want to work.

If you have a great idea, you could, of course, also found your own company. I am a friend of entrepreneurship, but be sure you know what you are doing before you invest all your life's savings into a company.

If you think SETI is a worthwhile endeavor and want to support it but don't want to go to that level of commitment, there are other ways for you to get involved, which don't require you to work full time on it.

One of the easiest ways to support SETI is to run the *SETI@home* App on your computer. Download *BOINC*[24], install, add *SETI@home* and you're done. Now every time you are not using your PC, *SETI@home* will search for signals from ETI. It's that simple. Consider though, that your PC has to do a lot of work. So this might influence its lifetime or your electricity bill, for instance.

If you know (or want to learn) how to code and analyze data, you can put your skills to the test with real data gathered by *Breakthrough Listen*. More info can be found on the website of the *Berkeley SETI Research Center*[25].

If you want to build your own small radio telescope and be part of a global network that searches for ETI signals (yes, you can actually do that), then take a look at the *SETI League*[26] and its *Project Argus*.

And lastly, if you don't want to work on SETI at all but still want to support the cause, you can always donate money to related institutions. The most prominent example is probably the *SETI Institute*[27]. But you could also give to space advocacy organizations like *The Planetary Society*[28].

Also, take a look at the section on interesting resources for more inspiration.

These are just some suggestions for how you could get involved. There are, of course, still other ways you could support SETI. Just choose whatever feels right to you.

Epilogue – So, are we alone?

There have been made many arguments against SETI. First, there were no other planets. Then the occurrence of life is too hard and just happened here. Then life appears regularly but doesn't evolve to multicellular life because it's too hard. Then life doesn't develop intelligence. Then that intelligence doesn't develop science. Then it doesn't develop technology. These, to me, are all just symptoms of the "center of the universe syndrome." The need to feel somehow special.

Humans have thought they are so special throughout their history. Until we found out that we're not. The Earth is not the center of the universe. The sun is not the center of the universe. And now that not even our galaxy is the center of the universe we have to be special in some other way. Now life is special, or at least intelligence or technology. I hope SETI will show us in the near future that we are special, but surely not alone.

It is my hope that you found this book to be interesting, useful and that maybe I could even inspire you to take action to support the cause of searching for life in the universe in some way.

Acknowledgments

This research has made use of the NASA Exoplanet Archive, which is operated by the California Institute of Technology, under contract with the National Aeronautics and Space Administration under the Exoplanet Exploration Program.

Interesting resources

I intentionally didn't put every resource I used for my book research in the following list. Instead, it's supposed to be a list of interesting links and resources to start you on your journey if you want to dive deeper.

I've purposefully only included free resources, with the only exceptions being the original *Nature* article by Cocconi and Morrison and the book *SETI 2020*, due to their importance to the subject. That doesn't mean that I think there are no good books about SETI out there. There are, and I encourage you to read them. But you can easily find them in the bookstore of your choice.

I wanted to make it as easy as possible for you to take the next step on your SETI journey. All you need to access the following resources is an internet connection.

General

- Google (or the search engine of your choice) is always a great way to start any research. https://www.google.com/
- Wikipedia is usually a great way to get a quick overview of any topic. https://www.wikipedia.org/

SETI (Search for Extraterrestrial Intelligence)

- "Searching for Interstellar Communications" (Nature article by Cocconi and Morrison that started it all). Original article (paid): https://www.nature.com/articles/184844a0 ; reproduction without mathematics (free): http://www.bigear.org/vol1no1/interste.htm
- Cyclops Report (the influential NASA study from the 1970s, some even call it the SETI bible). https://ntrs.nasa.gov/archive/nasa/casi.ntrs.nasa.gov/19730010095.pdf
- SETI 2020 (the result of a three-year effort by the SETI Institute to map out the future of SETI, kind of like a newer version of the Cyclops Report). https://www.amazon.com/SETI-2020-Roadmap-Extraterrestrial-Intelligence/dp/0966633539/
- An introduction to optical SETI by Stuart Kingsley. http://www.coseti.org/opticals.htm
- An introduction to optical SETI by Princeton University https://www.princeton.edu/~willman/observatory/oseti/oseti.html
- Centauri Dreams article about new All Sky All the Time Laser SETI project https://www.centauri-dreams.org/2017/07/14/laser-seti-all-sky-all-the-time/
- SETI Institute website. https://www.seti.org/
- SETI Institute FAQs. https://www.seti.org/faq
- Berkeley SETI Research Center website. https://seti.berkeley.edu/
- Berkeley SETI Research Center FAQs. https://seti.berkeley.edu/FAQ.html
- BOINC (the program necessary to run SETI@home). https://boinc.berkeley.edu/
- SETI@home About page. https://setiathome.berkeley.edu/sah_about.php
- SETI@home data activity (cool map of data

transmissions between Berkeley's servers and client PCs). https://setiathome.berkeley.edu/kiosk/
- SETI League, Inc. (home of Project Argus). http://www.setileague.org/
- All things SETI on The Planetary Society's website. http://seti.planetary.org
- Rio Scale (scale for assessing the importance of a candidate SETI signal). Official site by the IAA: http://avsport.org/IAA/rioscale.htm ; paper on revising the Rio Scale in "International Journal of Astrobiology" (2018): https://www.cambridge.org/core/journals/international-journal-of-astrobiology/article/rio-20-revising-the-rio-scale-for-seti-detections/DF9D6EABEA7D8D84999234BCFB3FADB4#

SETA (Search for Extraterrestrial Artifacts)

- SETV entry in David Darling's Encyclopedia. http://www.daviddarling.info/encyclopedia/S/SETV.html
- List of publications by Robert Freitas on the subject. http://www.rfreitas.com/AstroPubls.htm ; I suggest you start with "The Search for Extraterrestrial Artifacts (SETA)" http://www.rfreitas.com/Astro/SETAJBISNov1983.htm

METI (Messaging to Extraterrestrial Intelligence)

- "Making a Case for METI" guest editorial on SETI League homepage. http://www.setileague.org/editor/meti.htm
- "Rationale for METI" paper by Alexander Zaitsev. https://arxiv.org/abs/1105.0910
- METI International website. http://meti.org/
- Breakthrough Message website. https://breakthroughinitiatives.org/initiative/2

- San Marino Scale (scale for assessing the potential impact of a METI signal). http://www.setileague.org/iaaseti/smiscale.htm

Exoplanets

- Wonderful NASA website with all about exoplanets, presented in an easy to understand manner. https://exoplanets.nasa.gov/
- NASA App to take a closer look at what already found exoplanets might look like. https://eyes.jpl.nasa.gov/eyes-on-exoplanets.html

Fermi paradox

- A good article by Wait But Why on the Fermi paradox https://waitbutwhy.com/2014/05/fermi-paradox.html

Post-detection protocol

- "Concerning Activities Following the Detection of Extraterrestrial Intelligence" (what should be done after a SETI candidate signal is found. Officially adopted by the IAA but not legally binding to anyone). https://www.seti.org/protocols-eti-signal-detection

Telescopes

- Arecibo radio telescope. http://www.naic.edu/ao/landing
- Allen Telescope Array (ATA). https://www.seti.org/ata
- Green Bank Telescope (GBT). https://greenbankobservatory.org/
- James Webb Space Telescope (JWST). https://www.jwst.nasa.gov/

- Kepler Space Telescope. https://www.nasa.gov/mission_pages/kepler/main/index.html
- Low Frequency Array (LOFAR). http://www.lofar.org/
- Parkes radio telescope. https://www.parkes.atnf.csiro.au/
- Square Kilometre Array (SKA). https://www.skatelescope.org/
- Transiting Exoplanet Survey Satellite (TESS). https://www.nasa.gov/tess-transiting-exoplanet-survey-satellite
- Wide Field Infrared Survey Telescope (WFIRST). https://www.nasa.gov/wfirst

References

As this is not an academic paper, I have not referenced everything in this book. I used references to give credit where appropriate. Furthermore, I used references where it might be helpful to the reader if he or she wants to take a closer look at a topic or specific paper mentioned.

1. https://ntrs.nasa.gov/archive/nasa/casi.ntrs.nasa.gov/19730010095.pdf
2. https://arxiv.org/abs/1609.00330
3. ref https://arxiv.org/abs/1609.00330
4. https://arxiv.org/abs/1712.06639
5. http://www.rfreitas.com/Astro/TheCaseForInterstellarProbes1983.htm
6. Wikipedia contributors. (2018, July 14). Ufology. In Wikipedia, The Free Encyclopedia. Retrieved July 25, 2018, from https://en.wikipedia.org/w/index.php?title=Ufology&oldid=850217326
7. Wikipedia contributors. (2018, May 2). Sagan standard. In Wikipedia, The Free Encyclopedia. Retrieved July 25, 2018, from https://en.wikipedia.org/w/index.php?title=Sagan_standard&oldid=839363008
8. https://arxiv.org/abs/0805.2429
9. https://arxiv.org/abs/1203.2847
10. https://exoplanets.nasa.gov/5-ways-to-find-a-planet/
11. Wikipedia contributors. (2018, July 26). Biosignature. In Wikipedia, The Free Encyclopedia. Retrieved August 1, 2018, from https://en.wikipedia.org/w/index.php?title=Biosignature&oldid=852003565
12. https://mars.nasa.gov/mars-exploration/timeline/
13. https://arxiv.org/abs/1105.0910?context=physics

14. https://en.wikipedia.org/wiki/Active_SETI#Realized_projects
15. https://en.wikipedia.org/wiki/Voyager_Golden_Record
16. http://avsport.org/IAA/IAC07-A4.2.04.pdf
17. Wikipedia contributors. (2018, June 7). Multicellular organism. In Wikipedia, The Free Encyclopedia. Retrieved August 7, 2018, from https://en.wikipedia.org/w/index.php?title=Multicellular_organism&oldid=844799989
18. https://en.wikipedia.org/wiki/Technological_singularity
19. http://www.setileague.org/iaaseti/rioscale.htm
20. http://www.setileague.org/articles/bigear.htm
21. http://www.setileague.org/argus/
22. http://breakthroughinitiatives.org/about
23. http://breakthroughinitiatives.org/initiative/5
24. https://boinc.berkeley.edu/download.php
25. https://seti.berkeley.edu/listen/data.html
26. http://www.setileague.org/
27. https://www.seti.org/
28. http://www.planetary.org/

www.ingramcontent.com/pod-product-compliance
Lightning Source LLC
Chambersburg PA
CBHW031923240526
45464CB00022B/647